D1083415

RENEWALS: 691-4574

DATE DUE

MAR 12			
MAR 1 6			
JUL 1 9			
JUL 2 1			
JUL 2 8			
APR -9			
APR 2 8			
SEP 2 8			
MAR 07			
DEC 1 7 2008			

Demco, Inc. 38-293

SCIENCE, REASON AND RELIGION

SCIENCE, REASON & RELIGION

DEREK STANESBY

WITHDRAWN
UTSA Libraries
At San Antonio

CROOM HELM
London • Sydney • Dover, New Hampshire

©1985 D.M. Stanesby
Croom Helm Ltd, Provident House, Burrell Row,
Beckenham, Kent BR3 1AT
Croom Helm Australia Pty Ltd, First Floor,
139 King Street, Sydney, NSW 2001, Australia

British Library Cataloguing in Publication Data

Stanesby, Derek
    Science, reason and religion.
    I. Title
    215   BL240.2
    ISBN 0-7099-3360-6

Croom Helm, 51 Washington Street,
Dover, New Hampshire, 03820 USA

Cataloging in Publication Data applied for.

WITHDRAWN

LIBRARY
The University of Texas
At San Antonio

Printed and bound in Great Britain

CONTENTS

Preface

Acknowledgements

Introduction   1

Notes   5

1 The Authority of the Senses   6

    The Empirical Tradition: Positivism   6

    The Received View   10

    The Verification Theory of Meaning   13

    The Development of the Received View   18

    Induction   20

    Metaphysics and Religion   24

    The Failure of the Verification Principle   28

    The Demise of the Received View   31

    The Theological Response   35

    Retrospect   45

  Notes   46

2 The Retreat from Authority   48

  Karl Popper   48

  The Evolution of Persons   52

    The Evolutionary Continuum   52

    Emergence and Reduction   53

    Darwinism and Natural Selection   53

    Indeterminacy and Plastic Control   55

    The Emergence of Consciousness   56

    Problem Solving   56

    The Emergence of Language — the Genetic Factors   59

    The Emergence of Language — the Cultural Factors:

    an Evolutionary Sequence   60

    Linguistic Subjectivism   63

    Learning Language by Trial and Error   63

    Language, Plastic Control and Feedback   64

    Conclusion   64

  The Evolution of Knowledge   65

    Evolutionary Epistemology   65

The Popperian Account of the Growth of Knowledge   67
Demarcation   74
The Myth of Induction   79
Induction — The Psychology of the Situation   81
Induction — the Logic of the Situation   84
Truth and Verisimilitude   90
Induction Again?   96
The Implications for Religion   99
The Spirit of Enquiry   99
Metaphysics and Demarcation   102
The Retreat from Authority   108
Open and Closed Societies   115
The Open Universe   122
Evolution   127
Postscript   132
Notes   134

3 The Retreat to Irrationality   136
The Rationale of Discovery   136
T.S. Kuhn   138
P.K. Feyerabend   148
The Fundamental Flaw of the *Weltanschauungen* Analyses   156
Ludwig Wittgenstein   161
Paradigms, Forms of Life and Communities   167
Wittgenstein and Religious Belief   170
Religion without Explanation — D.Z. Phillips   173
Meaning and Existence   178
The Slippery Subject of Truth   183
Notes   186
4 Science, Religion and Rationality   189
Notes   196

References   198
Index   204

Beware lest any man spoil you through philosophy and vain deceit.   (Saint Paul to the Colossians 2.8)

PREFACE

The relationship between science and religion is one that has received increasing attention in recent years, and the attempt to exploit some of the implications of molecular biology and the new physics has given it fresh momentum. However, much of this work is misguided and philosophically flawed because the term 'science' is used uncritically, often with a deference combining reverence with suspicion, as if it were obvious to all what it is that distinguishes science from non-science.

Some writers attempt to establish theology as a science, thus abolishing the distinction; whilst others hanker after an all-embracing synthesis of natural science and dogmatic theology after the manner of Aquinas; and there are those, following Teilhard de Chardin, who reckon that by extending the boundaries of science we enter the realm of religion; some with modern cosmological theory in mind even suggest that science is the new religion. Most common of all are those who speak of two levels of truth, scientific and religious; or who, bowing to the authority of science, happily relegate religion to irrational faith.

An attempt is made in this book to evaluate critically the notion of science, to examine some of the assumptions implicit in its use, and to suggest that the links between science and religion are to be discovered at a much deeper, philosophical level. It is very much a ground-clearing exercise. But the ground must be cleared, our presuppositions examined, and implicit epistemologies exposed to criticism.

Such is the task of philosophy, and it is of crucial importance. Philosophy is not a detached, esoteric pastime for a gifted and leisured elite; it is fundamental to all our thinking. Philosophy affects the lives of ordinary people, for all human institutions are dependent on an epistemology of one sort or another. The danger today is that the sort of intellectual enquiry which is characteristic of philosophy is being sacrificed to the gods of pragmatism and productivity. Although many of the questions of philosophy are ongoing, some problems can be solved, and the exercise can be stimulating and exciting. But it involves patience and persistence, a constant wrestling with problems, and an acquaintance with the strivings and attempts of others, particularly of the intellectual giants of the past. Philosophy has an indispensable history.

Preface

Consequently a critical account is offered of the main movements in philosophy of science this century, of their influence on epistemology in general and on religious thought in particular. The powerful and all-pervading influence of positivism, and its offspring relativism, is examined, and hitherto unexplored implications of Karl Popper's philosophy of science for theology suggested. Authoritarian claims to truth and certainty in science are rejected, and the fallibilism of scientific knowledge is extended into theological thought, so replacing fixed and eternal dogma with tentative knowledge, always open to criticism and amendment. This is the area in which science and religion can find common ground and thus remove from man's mind the shackles of authoritarian dogmatism which stifle and enslave the human spirit with such disastrous consequences.

Such are the ingredients of this attempt to explore the implications of philosophy of science for philosophy of religion. Hopefully it may provide a much needed turning point for the philosophy of religion, and help to put theology back on the larger intellectual map where it belongs.

I would like to record my gratitude to some of the people who have inspired and supported me in this task; to Professor Raymond Plant for some useful and stimulating early discussions; to Dr David Lamb for his later supervision of this work for a doctoral thesis, and for his kindness and many helpful suggestions; to my friend Mr T.A. Greenslade for carefully typing the manuscript and for his astute and invaluable criticism; and above all to my wife Christine, for her unflagging support and encouragement. I would also like to thank the people of St Chad, Ladybarn, for their support and for relieving me of so many parochial duties whilst I was writing this. All of which goes to show that the writing of books is a corporate enterprise.

Finally I would like to dedicate this book to all parish priests, in the hope that in some remote but significant way it may give them courage to persist and help them to discern good philosophy from vain deceit.

Derek Stanesby
Ladybarn, Manchester

ACKNOWLEDGEMENTS

For permission to include extracts from the undermentioned works acknowledgement is made to the following copyright holders and publishers:
Hutchinson, London, for quotations from K.R. Popper, *The Logic of Scientific Discovery* (2nd edn, 1959, revised 1968); *Quantum Theory and the Schism in Physics* (1982); *The Open Universe* (1982); *Realism and the Aim of Science* (1983). Oxford University Press, London, for quotations from K.R. Popper, *Objective Knowledge* (1972, rev. edn 1979). Routledge and Kegan Paul plc, London, and Princeton University Press, Princeton, for quotations from K.R. Popper, *The Open Society and its Enemies* (5th edn, 1945, revised 1966, last reprint 1984). Routledge and Kegan Paul PLC, London, for quotations from K.R. Popper, *Conjectures and Refutations* (4th edn, 1963, revised 1972, last reprint 1984).

INTRODUCTION

In the Middle Ages theology was described as the 'Queen of the Sciences', that is, the highest and most authoritative form of knowledge. All rational enquiry had to conform to the canons of theological thought. The knowledge of God surpassed all other knowledge, and there was a sense in which all knowledge was subservient to the revealed truth of God, systematised by theologians and given the imprimatur of the Church. The religious view of the world dominated all thinking, and whenever there were clashes the religious view won the day.

This is not to say that there were clear and separately established disciplines, for example of religion, or theology and science.[1] Although Christian natural theology formed the seed bed for a later secular natural science, theological and empirical speculations were intimately related and interdependent. Christian theology was certainly not hostile to the spirit of empirical enquiry — in fact, the one encouraged the other; but because of the conception of rationality that held sway, empirical questions were contained within the overall metaphysical view of the day. If there was conflict it was between new ideas and old, rather than between religion on the one hand and science on the other.[2]

An earlier example of such conflict comes from fifth-century Athens, from which Anaxagoras was expelled for describing the sun as a red-hot stone. His views were contrary to the prevailing view of the world. But this was no clash between science and religion, because there were no recognised boundaries between them. It is an early example of the clash beween new ideas and old. It is only with hindsight that we discern a scientific spirit of enquiry emerging among the Ionian philosophers.

The great divide between science and religion was heralded by the Copernican revolution of the seventeenth century. From the time of Galileo onwards natural science, as a source of knowledge, developed with a life of its own. Although scientists such as Galileo and Newton were in the main religious men who subscribed to the beliefs and practices of the Church, natural science became independent of religion and of theological enquiry. Nevertheless, the religious view, particularly that resulting from Christian natural theology, continued to assert a strong influence over men's minds. This was as true of Einstein in the

1

twentieth century as it was of Newton in the seventeenth century and of Kepler and Copernicus before him.[3]

Today natural science rules as queen over all and is commonly accepted as the supreme source of all knowledge. One of the most influential thinkers of the twentieth century, Bertrand Russell, rooted his diverse and wide-ranging philosophy in such an assumption; and Karl Popper sees science as providing the best example of the growth of knowledge, therefore it is only through a study of the nature of scientific knowledge that one of philosophy's central concerns, epistemology, can be pursued. The tables have been turned. Contemporary religious thinkers now tend to take the authority of science for granted and they try to match their theology to the prevailing Western scientific tradition.

It is part of the task of philosophy to examine assumptions and presuppositions which underlie knowledge claims, and which underpin the institutions constructed upon them. Such philosophical enquiry has been directed to both science and religion. Religion has tended to come off by far the worse; it is found to be on very shaky ground when under the microscope of philosophical scrutiny. Certainly the British empirical tradition, from Locke to Hume to twentieth-century positivism, has been highly destructive of the claims of religion, whilst at the same time elevating natural science as the paradigm of all rationality.

But has Western science intellectually eclipsed religion as a rational activity? Religious practices continue; people still pray and worship and profess religious beliefs. And yet it is often asserted that contemporary natural science has relegated all religious belief and practice to the realm of pre-scientific ignorance and superstition. For this reason religion has been very much on the defensive. Religious thinkers have increasingly tended to accept uncritically the prevailing philosophical fashion and have attempted to use its methods in order to provide a rationale for religious belief and practice. This is as true of twentieth-century reductionism as it was of nineteenth-century Modernism. More recently, as a reaction to positivist assertions of the meaninglessness of metaphysics, and therefore of all religious talk, a whole generation of philosophers of religion, following Wittgenstein, has devoted its efforts to the examination of religious language. But such attempted defences of religion come from within the religious tradition; they are very much defences of something under attack. The battle is fought on terms dictated by the critics. Passing references, often of a most superficial and patronising nature, are made to science and its knowledge claims, but scientific activity is generally viewed as a well-established and philosophically sound enterprise which is immune to the manifold maladies with which the religious apologist has to contend.

However, there has developed in recent years an increasing interest in the philosophical examination of scientific thinking and practice. The philosophy of science is now one of the major areas of philosophical interest, as our subsequent discussion will indicate. The results of such enquiries have revealed widely differing accounts of the practice and rationale of science, and it would appear that in many respects the philosophical assumptions and presuppositions on which many have claimed that science is well based are as shaky and questionable as those on which religious beliefs and practice rest.

The purpose of this book is not to attempt to offer a philosophical defence of religion, nor to contribute to the age-old debate about the existence of God, nor to secure religion on sound, incontrovertible philosophical principles in order to provide it with a worthy intellectual status. Still less do we wish to make the misguided attempt to rehabilitate religion as a sort of science. Rather, we wish to examine the intellectual basis of science in order to demonstrate that the philosophical problems it throws up have much in common with those at the philosophical roots of religion; that the metaphysical assumptions of one are germane to the rationality of the other. Indeed, our discussion will be very much concerned with the whole notion of rationality and the various attempts that have been made to establish the canons on which rational thought is based. Our main concern will be a critical examination of some of the attempts to characterise science, and to identify the philosophical implications for religion that emerge from such philosophies of science. We will attempt to demonstrate the continuity and interdependence of philosophical ideas in science and religion; to show that common lines of philosophical thought run through both; that the fundamental questions are philosophical; that philosophy matters; above all to show that philosophy matters for two of man's greatest concerns: with God and with the world.

Both science and religion are human activities; each contributes to the characterisation of civilised man. Each has a profound bearing on human life and aspirations and achievements. They use a shared human language derived from common origins. Each begins at the same roots of puzzlement and enquiry, of wonder and awe, and the desire to find out how and why the world is. The fundamental concern of each is to grapple with the world and the place of man in it; to make some sort of sense of life; to discern some sort of order in the world. The one provides the assumptions and motivation for the other.

And yet science and theology tend to go their own separate ways. University faculties of science and theology seldom relate to one another. A person may go through life steeped in either scientific or theological learning and never attempt to relate, at least in any depth, his thought and discipline to the other. Theology and science are carefully segregated, few cross references are found in books. The one thing that scientists and theologians do share is a profound ignorance of each other's discipline. From the scientist's point of view theology is not empirically based and is therefore irrelevant to science. From the theologian's point of view science is not a part of the Biblically based revelation and is therefore irrelevant to his theology. Each searches for, and finds, within his own subject, authority. Although he allows criticism from within his own chosen field, criticism from without is either not heard or considered as an illegal, unwarranted trespass on his own chosen field.

There is no one philosophy of science, any more than there is one philosophy of religion. Although many of the questions and arguments are perennial, each is a continually developing subject. This is particularly so in the philosophy of science in which there has been considerable movement of thought in the middle years of this century. We have selected three of the most important movements which we label *positivism*, *Popperianism* and *relativism*, and have indicated their implications for the philosophy of religion. Consistent with the view that an understanding of the philosophy of science is fundamental to the philosophy of religion, we have attempted to offer fairly detailed and comprehensive accounts of the philosophical positions concerned. In the first instance an attempt has been made to give a clear and straightforward exposition, using where possible the words of the philosophers themselves,[4] reserving critical comment to the end of each exposition, followed by the implications for religion. If philosophy of science is important for philosophy of religion, then it is important for the philosopher of religion to understand the philosophy of science on its own terms. This is why we have attempted to view the philosophy of religion from within the philosophy of science, rather than the other way round.

Many of the arguments, like so much in philosophy, are in the end inconclusive. But the *arguments* are of supreme importance, for philosophy affects the way we live our lives, the way we treat each other, the values we hold, and the sort of institutions we construct. In a science-dominated age men hardly need to be persuaded that science is important (in terms of its potential for both good and evil); but we do need to be reminded that religion is important (again in terms of its power for good or evil), because it takes into account the incredible complexity and limitations and mystery of human nature, without which men wonder who they are.

Notes

1. On the whole, whilst aiming at clarity, we will avoid the fruitless attempt at definition. The terms 'theology' and 'religion' will be used fairly loosely, sometimes being interchangeable. In general 'religion' will be used as the wider term, embracing the beliefs and practices of people; and 'theology' will refer to the attempt to systematise forms of religious belief into a rational and coherent pattern.

2. The dependence of modern natural science on the presuppositions originally provided by Christian natural theology have been well documented. Following Whitehead's suggestion in *Science and the Modern World* (1938), Hookyaas (1972) and Jaki (1978), among others, have produced detailed and well-argued accounts.

3. Particularly in terms of the harmony and intelligibility of the universe, a profoundly mystical assumption.

4. This is particularly so in the case of Popper, who is quoted liberally simply because his words are better than any summary account. As Anthony Quinton has remarked, 'A conspicuous virtue of his work is the undeviating clarity and definiteness of the language in which it is expressed . . . he has a moral passion for rational intelligibility' (Quinton, 1982, p. 293).

1   THE AUTHORITY OF THE SENSES

The Empirical Tradition: Positivism

What is the basis of the contemporary esteem of science, its appeal and its authority? For all that there has been an increasing element of disillusion regarding science and its application in the field of human affairs, a disillusion generated by some of the dubious and deleterious effects of applied science and technology, there is no doubt about the hold that modern science has over men's minds and the control it exercises over their lives. Indeed, to the extent that people have become disenchanted with science, that very disenchantment is based on fear; fear of the power and authority and influence of science.

To what is this authority and appeal attributed? It is commonly held that science derives its power and authority from its empirical method; a method that comprises a sure and reliable inference from observation and experiment. Its authority derives from our senses. On the objective base of our sense experience our scientific knowledge is constructed. The corollary to this assumption is that knowledge that is not based in some way on sense experience is rejected as ill founded and illusory. Hence the rejection of religion by a large majority in the science-dominated Western world.[1]

Francis Bacon, in the early seventeenth century, was one of the first to attempt to articulate what the method of modern natural science is, and he became the forerunner of the British empirical tradition. Although he appears to have accepted the doctrines of Christianity as true, and he allowed room for natural theology, Bacon emphasised the distinction between knowledge inspired by divine revelation and knowledge arising from the senses. Bacon was very much concerned with ameliorating man's lot on earth, and for him that utilitarian aim was best achieved not by detached, speculative thought but by the collection of facts through organised and systematic observation and deriving theories from them. He believed that only such a systematic and detailed experimental method would give man new knowledge of the natural world. This new knowledge, derived through the senses, would give man power to transform human life for the better. Bacon's anti-scholasticism led him to reject traditional syllogistic logic as a means of empirical discovery, and he insisted that if we are to understand nature we must consult nature and not the writings of Aristotle.

Bacon inherited from Plato and Aristotle the notion that the mind is tainted by error and false belief, but unlike his Greek predecessors he held that the source of true knowledge is Nature itself, which does not lie. Consequently the mind has to be purged from all anticipations, conjectures and guesses which are the source of error and impurity. The scientist is thus urged to observe the world around him in order to prepare his mind for the unbiased interpretations of nature. Nature, according to Bacon, 'bears the signatures of God, and it is these, the true forms of things, which are the goal of natural philosophy, and not the false images imposed on things by man's mind' (Hesse, 1964, p. 143). Thus Bacon replaced the authority of religious or philosophical conviction, that is the external authority of revelation or the internal authority of reason, with the authority of the senses.

Here we have the roots of the empirical tradition which has had such a powerful influence on British philosophy. Such is the appeal and authority of observation or sense experience that philosophers from Locke to Russell to members of the Vienna Circle have been preoccupied with establishing and refining the empirical method as the only intellectually respectable account of the way knowledge is advanced.

The central concern in this endeavour consists in a scrutiny of the nature and structure of scientific theories in all their diverse roles in the scientific enterprise. Scientific knowledge finds its expression and application in the great variety of theories that is produced in the name of science. Popper ascribes a vicarious role to the theories that men produce. They play a decisive part in the evolutionary scheme and on them our survival depends. Rational man has emerged from the realms of biological evolution armed with theories with which he can test his environment and probe into the unknown. His theories provide him with undreamed of power and control in the world in which he finds himself. The centrality of theories to science in both its practical and philosophical aspects is well attested by recent history. As Suppe observes, 'the last century has provided science with some of its most spectacular, controversial, and revolutionary theoretical episodes in all branches of science — physical, biological and social' (Suppe, 1977, p. 3). We have only to consider Relativity and Quantum theory in physics, Darwinism in biology, Marxism in the political and economic realm, and the work of Freud, Adler and Jung in psychology to be reminded of the profound and far-reaching effect of theories on the affairs of men.

It was, then, to an examination of the theories of science that philosophers turned their attention. And because modern natural science flourished once it had been freed from its Aristotelian strictures it appeared that the only satisfactory rationale of scientific theorising lay in its empirical method. Scientific knowledge is based on what is given through observation and sense experience.

It was with this basic assumption concerning the crucial role of sense experience that philosophers analysed the theories of natural science. The greatest successes in natural science in the nineteenth century were in the realm of physics, and so it was that physics became the paradigm of science. The philosophy of science in the last century has in reality been the philosophy of physics. The implicit assumption contained in this approach is that all genuine science can be reduced to physics, or that a subject is scientific to the extent that it conforms to the discipline of physics. The appeal of the physical model for modern science is precisely the appeal of the mathematical model for Greek science: that is the appeal of precision and certainty, and therefore of authority.

The positive acquisition of knowledge via one's sensory equipment thus became the hallmark of a theory of knowledge labelled *positivism*. If we have identified the roots of positivism in the writings of Francis Bacon, it was the British empiricist school of the seventeenth and eighteenth centuries that attempted to develop these ideas into a systematic theory of knowledge. Auguste Comte in nineteenth-century France attempted to develop a positive sociology and to provide a truly scientific basis for the re-organisation of society. He shared Bacon's optimism about the benefits of a positive approach to science for humanity. For Comte, his positivistic programme even formed the basis of a new religion in which the worship of Humanity replaced the worship of God. J.S. Mill paralleled Comte's efforts in England with special regard for the logic of science and a general scepticism towards religion. Herbert Spencer developed important and influential ideas on evolutionary theory, emphasising its positive, empirical basis.

This positivistic programme had a hard-headed and commonsensical appeal. Knowledge is based on empirical facts, to be accepted for what they are without going beyond what is given or laid down. The extravagancies of metaphysics and theology which attempted to go beyond the world of observation into first causes and ultimate ends were ruled out of this positivistic programme. All genuine knowledge is contained within the boundaries of natural science.

But it was in the German scientific establishment of the nineteenth century that positivism developed, and it achieved its most powerful formulation in the writings of those mathematicians, scientists and philosophers who formed the nucleus of the Vienna Circle in the first decades of the twentieth century. What began as an attempt to establish the philosophical foundations of scientific knowledge flowered into a full-blown epistemology. Long after the general epistemological claims of Logical Positivism had been discredited and proved inadequate, it still exerted a powerful influence in the specialised area of the philosophy of science.

The reasons for the survival of positivism as the philosophy of science are both interesting and instructive. Criticisms of positivism came from within and without the scientific establishment. It was maintained by many that not all empirical knowledge is like scientific knowledge,[2] and therefore positivism was rejected as a general epistemology, but not questioned as the only satisfactory analysis of scientific knowledge. This attitude is still exhibited by many apologists of religion whose defence of religious language and theological knowledge is that it is not like scientific knowledge. What these philosophers of religion failed to perceive, doubtless through their lack of acquaintance with science, was that the positivist programme for science itself is inadequate. The philosophy of religion in the twentieth century has tended to be a weak and defensive philosophy, impressed by the power and authority of science, but based on a somewhat naive view of the philosophy of science.

It is often maintained that Logical Positivism emerged as a reaction to the metaphysical excesses of the neo-Hegelians of the late nineteenth century. Russell and Moore in Britain succeeded in launching a destructive, critical crusade against their Idealist mentors, McTaggart, Bradley and others, who had sought to explain Reality in terms of extreme metaphysical abstraction, such as Entelechy and the Absolute, which has no empirical or factual foundation. There is no doubt that the members of the Vienna Circle were inspired by what they saw as the prevailing spirit of enlightenment in Vienna and they were consequently fired with a crusading, evangelistic mission to eliminate metaphysics from every area of life. The vocabulary which they chose to further their cause, with its notions of 'cleansing' and 'clarification', had an almost puritanical ring. The purpose of their mission was to propagate a 'scientific world-conception' which would penetrate every form of life, 'in education, upbringing, architecture, and the shaping of economic and social life according to rational principles'. But the immediate goal of this endeavour was a 'unified science' in which

neatness and clarity are striven for, and dark distances and unfathomable depths rejected. In science there are no 'depths'; there is surface everywhere: all experience forms a complex network which cannot always be surveyed and can often be grasped only in parts. Everything is accessible to man: and man is the measure of all things . . . the scientific world-conception knows no *unsolvable riddle*. (from an unsigned pamphlet published by members of the Vienna Circle; in Neurath and Cohen, 1973, p. 306)

Doubtless the almost messianic fervour with which some positivists pursued their anti-metaphysical crusade contributed to the emergence of Logical Positivism, and non-scientists such as the young A.J. Ayer inherited this same desire to eliminate metaphysics from rational thought (Ayer, 1946). But its real origins, which we have traced back at least as far as Bacon, were inspired by a genuine and positive attempt to deal with foundational issues in the philosophy of science, and not simply with the negative programme of destroying metaphysics and religion. In any case, the overall reductionist programme which was the goal of the Logical Positivists depended on the fundamental reduction of natural science to statements about the given, that is the directly observable; and unless that initial reduction was clearly established there was little point in attempting to extend it to more contentious areas. It is to this positive analysis of natural science that we must now turn.

The Received View

Logical Positivism was originally a Germanic movement which evolved from the prevailing notions of scientific common sense in the German scientific community of the late nineteenth century. Frederick Suppe has identified three forms of the expression of this common-sense view, each designed to provide a philosophical basis for the developing classical physics of the day (Suppe, 1977, Ch. 1).

'Mechanistic materialism' viewed the world as a mechanical system, and science as the search after the mechanisms at work in this objective, material world. Empirical investigation yields knowledge of the mechanistic laws governing the workings of the world. Observation is central to this exercise; philosophical speculation and *a priori* knowledge have no place. Its chief spokesman, Ludwig Buchner, writes:

Science . . . gradually establishes the fact that macrocosmic and microcosmic existence obeys, in its origin, life, and decay, mechanical laws inherent in things themselves, discarding every kind of

super-naturalism in the exploration of natural events. There is no
force without matter; no matter without force. (quoted in Passmore,
1957, p. 34)

Mechanistic materialism was challenged on the grounds that it allowed
no place for the thinking subject in the growth of scientific knowledge.
Consequently a form of neo-Kantianism developed which was concerned less with immediate sensations and more with phenomenal structures. Kant had demonstrated the importance of the thinking subject in
his epistemology and so he established a link between sensibility and
understanding. There could be no neutral observer of phenomena. Just
as Copernicus had claimed that the apparent movement of the stars was
really in part the movement of the observer, so Kant suggested that our
observations of the phenomenal world were the product of the enquiring mind and the world itself. The consequence of this Copernican revolution in epistemology was to dethrone the observer as the direct
recipient of empirical or scientific knowledge. Scientific knowledge of
the world is not directly given, but rather revealed in the structure of
sensations, and the task of science is to discover the underlying structure of the phenomenal world. Knowledge of this structure is absolute
and consequently relativistic physics could have no place in this
common-sense view in which absolute space and time were fundamental
categories of experience.

As a further reaction to the earlier mechanistic materialism, and as
an alternative to the widely held neo-Kantianism, the views of Ernst
Mach gained increasing acceptance. Mach's neo-positivism allowed no
place for *a priori* elements in science. Consequently absolute space and
time were ruled out, and this proved crucial to the acceptance of relativity theory and in turn later invested Mach's views with a greater
degree of credibility amongst philosophers of science. But his opposition to atomic theory was not vindicated by later developments in
science, and neither was his expectation that physics would be
advanced through the physiology of the senses. Central to Mach's positive analysis of science are the sensations of the observer which he sees
as the key to all genuine scientific knowledge. Mach's sensationalism has
been labelled by Peter Alexander as 'epistemological atomism' because
it reduces scientific statements to basic 'atoms' of experience or sensation. According to this view, 'All statements which can be regarded as
the premises or conclusions of science can be analysed into, or are
reducible without remainder to, statements about these atoms and the
relations between them.' In other words, all genuine statements of

science must be empirically verifiable. Alexander comments that 'This account has been, and still is, one of the most influential of its kind, perhaps because it represents a determined attempt to apply an extreme empiricism in a complete and consistent way to science' (Alexander, 1963, pp. 1-2).

It was Mach who provided the link between Bacon and the earlier British empiricists Locke, Berkeley and Hume, and the later 'scientific' empiricism of the Logical Positivists. The enduring motive throughout this enterprise was to ground our knowledge on as firm a foundation as possible. Mach, who stands at the beginning of a new and important period in the philosophy of science, has had a lasting influence on the common-sense, accepted view of science. Alexander comments that Mach's account 'is one to be reckoned with; although many philosophers nowadays regard it with suspicion, many non-philosophers appear to regard it as fundamentally the only possible and obviously correct view' (Alexander, 1963, p. 3).

These three views of the nature of science were by no means mutually exclusive and they comprised the prevailing common-sense view of science, and therefore of rationality. They were construed to accommodate the classical physics of the nineteenth century, but all this was shattered by the epoch-making theories of Planck and Einstein in the realms of Quantum theory and Relativity theory. In fact it was part of Einstein's great achievement to introduce new concepts in physics which were contrary to accepted notions. Einstein in particular had been influenced by Mach's ideas, especially with regard to space and time, and his work in Relativity appeared to be philosophically grounded in a form of positivism. Although much of his work, particularly in General Relativity, appeared as a complex mathematical treatise, it suggested clearly empirical, testable consequences.

The philosophy of science that emerged as a result of the new physics of the early twentieth century was a weakened version of Mach's neo-positivism. In this respect the work of Reichenbach and Schlick was the most influential. The central problem for the philosophy of science was that of comprehending the role of theoretical, non-observable concepts in physical theory which Mach's extreme sensationalism disallowed. This became increasingly urgent with the growing use of complex mathematical constructions in the development of physical theory. The way out of this dilemma was to construe the theoretical and mathematical terms in physical theory as conventions or abbreviations for phenomenal descriptions. 'Mass', for instance, was operationally defined in terms of specific measurements on certain kinds of phenomena. Thus physical laws could be expressed in mathematical terms, but these terms would always be translated into phenomenal language and thus open to empirical verification. The work of Frege, Cantor and Russell, culminating in *Principia Mathematica* (Whitehead and Russell, 1910-13), explicated the logical basis of mathematics. If, as *Principia Mathematica* maintained, mathematics could be axiomatised in terms of logical theory, then mathematical statements of scientific laws could be interpreted as a sort of shorthand or abbreviation of phenomenal or observational descriptions.

The central tenet of positivism is retained in this treatment of scientific theory: that is, empirical verification is based on observation. Theoretical terms could always be reduced to observation terms, the reduction being accomplished through the medium of *correspondence rules*. These rules of interpretation are operational definitions which clearly specify the manner in which theoretical terms can be defined by phenomenal properties. A theoretical term such as 'mass' is so defined by experimental procedures that it retains its cognitive significance according to empiricist principles.

This is the view of the philosophy of science developed by members of the Vienna Circle, and it is aptly labelled *Logical Positivism* in that it attempts to relate the logical and experiential nature of the scientific enterprise. It is empiricist and positivist in that it admits of only knowledge from experience, which rests on what is immediately given; and it is logical in its method of reduction of scientific statements, step by step, to the given. Metaphysical entities and concepts were avoided by insisting that the only theoretical terms allowed were those that could be provided with correspondence rules to give them explicit phenomenal description. It is a view of the nature of scientific theories which emerged as a synthesis of the various philosophical, scientific, mathematical and logical developments and doctrines we have indicated.

This analysis of scientific theories as axiomatic calculi which are interpreted by means of correspondence rules is referred to as the *Received View on Theories*.[3] Suppe observes that for over thirty years this view exerted near total dominance over the philosophy of science. 'It is little exaggeration to say that virtually every significant result obtained in philosophy of science between the 1920's and 1950 either employed or tacitly assumed the Received View' (Suppe, 1977, p. 4).

The Verification Theory of Meaning

The Received View thus eliminated objectionable, non-empirical entities from scientific theories, and because science was seen by the Logical Positivists as the paradigm of rationality it was a natural move to extend this rationale to all areas of human discourse. Influenced by Russell's method of logical analysis as the new-found tool of philosophical clarification, and particularly by the early work of Wittgenstein, the Received View on scientific theories was extended into a general doctrine of cognitive significance.

The influence of Wittgenstein in this respect cannot be overestimated, and as A.M. Quinton remarks, the *Tractatus Logico-Philosophicus* (Wittgenstein, 1951) became the Bible of the logical analyst movement.

> Like other sacred texts, it combined prophetic fervour with sybilline obscurity in a way that invited and received many conflicting interpretations . . . it seemed that Wittgenstein, assuming the posture of a founder of a religion rather than the exponent of a philosophy was more unwilling than unable to make the task of understanding him an easy one. (Quinton, 1964, p. 535)

With the work of Frege and Russell in mind he tried to develop the doctrine of a logically perfect language.[4] If mathematics could be derived from basic, indisputable, logical concepts and axioms, then language ought to exhibit such a structure and this in turn would reveal the structure of knowledge itself. The principle of extensionality suggested that compound linguistic propositions could be analysed into basic, or elementary ones whose truth or falsity could be directly established. Hence the truth or falsity of a compound proposition was a truth function of its elementary constituents.

Wittgenstein, adopting this principle (3.318),[5] developed his picture theory of language in which he endeavoured to establish the relationship of language to the world. Logical analysis of a complex proposition would reveal a combination of elementary propositions which in turn could be further analysed into their component parts. These consisted of *names* or *signs* which are the simple building units of language. Here the analysis stops and we have a one to one relationship of language to the world (3.201, 3.202). The simple names picture the objects which are the ultimate constituents of the world: 'The name means the object. The object is its meaning' (3.203). An arrangement of objects constitutes a fact or a possible state of affairs (2.01, 4.031). 'The configuration of simple signs in the propositional sign corresponds to the configuration of objects in the state of affairs' (3.21). The distinction between a *state of affairs* and a *fact* allows Wittgenstein to distinguish between truth and falsity in meaningful language. A state of affairs is a logically possible fact (or arrangement of objects), whereas a fact is a state of affairs that happens to be the case. Hence a meaningful proposition is a logical picture of a state of affairs; it is true if it corresponds to an actual state of affairs (a fact) and false if it does not. Meaningless combinations of words, pseudo-propositions, violate the logic of language. 'Everything that can be said can be said clearly' (4.116) implies the corollary that what cannot be said clearly cannot be said significantly. Thus Wittgenstein attempted to demarcate sense from nonsense and concluded that only the propositions of natural science are legitimate and all metaphysical assertions have no meaning (6.53).

Howerever, the somewhat paradoxical closing remarks of the *Tractatus* carry the strong implication that what cannot be said significantly, that is the mystical, is of profound importance. The final statement of the *Tractatus*, 'Whereof one cannot speak, thereof one must be silent,' (7), is open to the same ambiguous interpretation, although the positivists took it as a total dismissal of metaphysical assertions from significant discourse as nonsensical, and maintained that there is nothing to be silent about. By his own strictures Wittgenstein in the *Tractatus* violated the logic of language, as he himself observes (6.54). The fact that he chose to speak rather than remain silent is a final vindication, if not of metaphysical language, then of a realm beyond language. There is no doubt that members of the Vienna Circle who looked for positive support from Wittgenstein in their anti-metaphysical endeavours were somewhat exasperated and perplexed when he addressed their meetings with poetry readings, or even occasionally by just whistling to them from his extraordinary repertoire of classical music (Bartley, 1974, p. 35).

However that may be, the *Tractatus* was taken as an important contribution to the anti-metaphysical armoury of the Logical Positivists. Although Wittgenstein insisted that his thesis was a logical one and that therefore it was not his task to give examples of elementary propositions or atomic facts, this deficiency was soon remedied by members of the Vienna Circle who identified the simple objects of the *Tractatus* with the objects of direct sense experience. Propositions referring to direct sense experience were labelled protocol sentences, and these constituted the basic statements of science which expressed indubitable facts about the world and imparted certain knowledge. In order to avoid the inherent metaphysical tendencies of ordinary language ingenious attempts were made at constructing a logically perfect, or protocol, language which would not only be the language of science but the language of all meaningful discourse. What had begun as an attempt to establish the rationale of science was extended to become the rationale of all linguistic assertions and therefore of all knowledge claims. Thus emerged the famous verifiability criterion of meaning expressed in the slogan 'the meaning of a term is the method of its verification'. Language that was used contrary to this dictum was dismissed as devoid of significance, and consequently a good deal of previously accepted discourse was rejected as metaphysical verbiage.

A distinction must be made between the *Verification Principle* which demarcates sense from nonsense, and the *Verification Criterion* which provides a method for establishing the meaning of a statement.[6] A.J. Ayer made several attempts to formulate a satisfactory Verification Principle, and thus legislate between sense and nonsense (Ayer, 1936 and 1946). But in either formulation empirical testability is essential for meaningful language. Strict Logical Positivism was initially concerned with the problem of verifying particular rather than general assertions. This was reduced to the problem of verifying observation or protocol language assertions. Two approaches were explored by different members of the Vienna Circle, the physicalist and the phenomenalist. These two views were considered as alternative ways of talking about the same thing.

The doctrine known as *physicalism* considered that language describing physical things is common to all language users, and therefore observations are inter-subjective and as such empirically verifiable according to the Verification Principle. The verification of such physicalist assertions was thus taken as unproblematic but it nevertheless contained a concealed doctrine of perceptual knowledge. If we agree about what we see then we are verifying our knowledge claims. Thus Carnap aimed at the development of one language for science, a physicalist language, purged of all metaphysical elements (Carnap, 1934).

A more sophisticated doctrine of perceptual knowledge involved the reduction of physicalist, or thing-language, to basic atoms of perceptual experience labelled sense-data. In this phenomenalistic approach the protocol language would be the sense-data language. The verification of such assertions was immediate and incorrigible because one's basic sensory experience was simply given, and therefore self-authenticating. These phenomenal descriptions of sense-data constituted the basic or protocol language. Such a phenomenalistic theory of perceptual knowledge was developed by Russell as early as 1912 and it has received its latest refinement from Ayer who, following the

American philosophers C.I. Lewis and Nelson Goodman, now prefers to refer to the logically neutral 'qualia' rather than to sense-data (Ayer, 1973).

Although phenomenalists like Ayer continued to court the reductionist thesis of perceptual knowledge, it was the less sophisticated physicalism that was embodied in the Received View as a doctrine of perceptual knowledge. The verification of physicalist assertions was taken as non-problematic; material things have observable properties for all to see.

The Verification Theory of Meaning attempted to extend a doctrine of science to a doctrine of language. Empirical verification, which was the foundation of the positivist view of science, became a prescription for all meaningful and sensible use of language. Verification became the password to knowledge. Observation terms are the basic units of language, and these are learned by ostensive definition as language is acquired in its early stages. All other terms are either theoretical terms which are reducible to observation terms, or logical terms which have a grammatical or syntactical function and are introduced by definition. Considering that the only legitimate use of language was reckoned to be scientific, it is not surprising that the theory of language acquisition also characterised the way in which science has developed. Science begins in true Baconian fashion by looking at the world and deriving from one's observations empirical generalisations. Later, theoretical terms are introduced by definition and laws are formulated. 'Thus science proceeds "upward" from particular facts to theoretical generalisations about phenomena, this upward process proceeding in an essentially Baconian fashion' (Suppe, 1977, p. 15).

How does such an account of the development of human language allow for the inherent metaphysical tendencies of ordinary language? What is the source of these impurities which so seductively mislead us in our daily discourse and which need exposing by logical analysis? The Logical Positivists offered an almost Marxist account in their talk of 'Fierce social and economic struggles' of the past and present, and they looked back to a golden age of language unsullied by the reactionary forces of religion.

The representatives of the scientific world-conception resolutely stand on the ground of simple human experience. They confidently approach the task of removing the metaphysical and theological debris of millennia. Or, as some would have it: returning, after a metaphysical interlude, to a unified picture of this world which had, in a sense, been at the basis of magical beliefs, free from theology, in the earliest times. (Neurath and Cohen, 1973, p. 317)

The Development of the Received View

Although the Logical Positivists of the Vienna Circle influenced philosophical thought well beyond the bounds of the philosophy of science, particularly through the influence of the Verification Principle, their prime concern derived from an interest in science and its foundations. Consequently they continued to refine the Received View by criticism in order to make it as sound a doctrine of science as possible. It soon became evident that the rigorous exclusion of metaphysical entities was not a simple task, and modifications had to be made to loosen up the theoretical components of scientific theories.

In its original formulation the Received View attached little importance to the theoretical apparatus of scientific theories, its main function being mathematical and as such shorthand for observable entities. But as the view developed under criticism this theoretical apparatus was given increasing status because it became apparent that non-observable or theoretical entities could not be completely reduced to their observable manifestations. The meaning of theoretical terms such as 'mass', 'force' or 'entropy', for instance, is not exhausted by observable properties. Further, dispositional terms such as 'fragile' could not be empirically defined by correspondence rules, but are clearly of cognitive significance. The problem here is basically that of the impossibility of exhaustive definition, every definition involving undefined and therefore theoretical terms. Hence the role of correspondence rules became less rigorous, it being sufficient to demonstrate that the theoretical system could be *partially* interpreted by correspondence rules. This represented a considerable shift in the positivists' doctrine of cognitive significance, a non-empirical element being admitted as essential in the formulation of scientific theories.

The debate concerning the nature of scientific theories took a number of turns as attempts were made to establish their status and cognitive significance. Mach's original sensationalist thesis attempted to reduce scientific explanation to description of phenomena because the concept of explanation contained a metaphysical component. According to this view science was descriptive rather than explanatory, the theories of science being conceived as instruments or tools for deriving one set of observation statements (predictions) from another set of observation statements (data). *Instrumentalism*, as it became labelled, avoided questions of the truth or falsity of theories by insisting that theories do not correspond with reality. The function of a scientific theory is simply that of a useful instrument for making predictions about phenomena. Although the Instrumentalist thesis can be traced back to the early formulations of the Received View, for example by Schlick, Ramsey and Waismann, Stephen Toulmin was still defending a form of Instrumentalism well into the 1950s (Toulmin, 1953).

However, the persistence of irreducible theoretical components in a scientific theory constituted an increasing embarrassment to positivists. Hempel described it as the 'Theoretician's Dilemma' (Hempel, 1965). His attempt to show that theories could be explanatory on the Received View was an admission that theoretical terms could not be eliminated from theories, but their status continued to be instrumental rather than realistic. The Theoretician's Dilemma arose from the acceptance of the necessity of theoretical terms but a denial that they have any meaning or reference to the real world.

The underlying question in the Instrumentalist/Realist controversies over the interpretation of theoretical language assertions was whether or not verification was a sufficient condition for obtaining knowledge about non-directly observable phenomena. This is a particularly pressing problem in the realm of particle physics, and it throws up questions, for instance, about the status and reality of electrons. Is an electron an instrumental fiction, or does it designate a real entity?

If *Instrumentalism* was an attempt to establish the status of scientific theories, then *Operationalism* was concerned with their cognitive significance. P.W. Bridgman in *The Logic of Modern Physics* attempted to give a satisfactory account of the significance of theoretical terms in a scientific theory by suggesting that correspondence rules be operationally defined. The concept of length, for instance, is determined by the operations used for measuring length. 'The concept is synonymous with the corresponding set of operations' (Bridgman, 1927, p. 5). It was in effect an elaboration of the Verification Criterion 'The meaning of a term is the method of its verification.' Meaning is method or use.[7] Operationalism constituted an extremely influential attempt to establish the meaning of scientific concepts in accordance with the logical empiricism of the Received View. Both Instrumentalism and Operationalism arose out of attempts to refine and develop the Received View on Theories according to the canons of positivism in which empirical verifiability was the central dogma.

Although the Logical Positivists' chief concern was that of providing a satisfactory analysis of scientific theories in which the various components of the theory could be reduced as closely as possible to observable entities, the emphasis on verification and confirmation of theories implied a cumulative view of the growth of scientific knowledge. It was recognised that older scientific theories are superseded by newer theories, but it was considered that the older, redundant theories are incorporated into the new theories. A positivist programme could not allow a once confirmed or verified theory to be abandoned or disconfirmed. Once-verified theories could not be assigned to the scrap heap. Old successes and new successes combine and augment one another. The authority of science could not be undermined by the admission that it had made mistakes.[8] This cumulative view of the growth of scientific knowledge involved a concealed doctrine of human progress, which in turn was inspired by the increasing power and control that scientific knowledge gave man.

Induction

The reaction against Aristotelian science, exemplified by Bacon, involved a shift from the deductive to the inductive method.[9] Bacon did not reject the deductive logic of the syllogism, but he recognised that it could only exhibit the consequences of what is already known, and could not lead to new empirical knowledge. He insisted that all new knowledge about the world must come from some form of induction from observed instances to general conclusions. According to this inductivist account of the growth of knowledge, the scientist carries out experiments and he meticulously records measured observations. Other scientists engaged in the same field of research accumulate similar data. As the body of experimental information grows, general features emerge which are formulated into hypotheses which fit all the observed facts and explain the causal connections between them. The hypothesis is confirmed by further observational evidence which supports it. Such verification of the hypothesis helps to establish it as a law of nature, which takes the form of a universal statement. Thus our knowledge grows by gradual accretion of such scientifically established 'laws'. Karl Pearson, a true nineteenth-century Baconian, maintained:

> The classification of facts, the recognition of their sequence and relative significance is the function of science . . . let us be quite sure that whenever we come across a conclusion in a scientific work which does not flow from the classification of facts, or which is not directly stated by the author to be an assumption, then we are dealing with bad science. (Pearson, 1936, pp. 11 and 14)

The inductivist account of the genesis of theories or laws from observational evidence allows for particular predictions to be established by deductive reasoning. For example, from the laws of planetary motion, which are established by induction, a particular solar eclipse could be predicted. Newton himself regarded the heliocentric system as a true description because it was deduced from the phenomena and 'rendered general by induction' (Hesse, 1961, p. 2).

The fundamental expression of the inductive method of reasoning by which a general law is inferred from accumulated observations of particular instances, referred to as 'induction by simple enumeration', takes the form A_1 is b, A_2 is b, A_3 is b, ... A_n is b, therefore all As are b. Some such inductive method is essential to the positivists' account of science, and consequently they endeavoured to establish a satisfactory inductive logic which in turn helped to characterise their rationale of science. The central epistemic problem in this endeavour consisted in attempting to demonstrate that inductive inferences, from a finite number of instances to a universal law, do provide knowledge.

The shadow of David Hume, the great empiricist philosopher, loomed over this endeavour. Although he never used the term 'induction' in this context, Hume discussed the problem in a classic passage in his *Enquiry* (Hume, 1951, Section 4). Hume's problem is that induction is incapable of *proving* the truth of a generalisation on the basis of examined instances. He argued that such 'inductive proof' was possible only if the premises of the argument contained some sort of known induction hypothesis concerning the uniformity of nature such as 'the future will be like the past'. But such an induction hypothesis is itself a contingent generalisation, which in turn is in need of justification or proof. Such justification could only be provided by induction, which in turn needed further inductive justification, and so on into an infinite regress. Proof, for Hume, involved logical entailment. No such logical entailment could be discerned in inductive inference. However one tried to justify inductive inference, one ended up with an inductive inference which itself needed justification.

Hume concluded that therefore science was irrational. No rational justification could be provided on which scientific knowledge could be constructed. However, despite his discovery of the irrational nature of inductive inference, Hume retained it for psychological reasons. Impressed by the success and predictive power of Newtonian physics in particular, and by scientific discovery in general, he maintained that we come to expect regularities as a result of constant repetition. I expect the sun to rise tomorrow because that expectation has been constantly satisfied in the past. I expect the billiard ball to move when struck because I have observed such behaviour many times before. Nevertheless, he maintained, my expectations in these instances, as in all cases of inductive inference, are purely psychological. They have no rational foundation. Russell summed up the problem in the parable of the chicken, who, accustomed to being fed each morning, expected the same treatment on Christmas Eve, only to have his head chopped off (Russell, 1952, p. 63).

Although the Logical Positivists accepted the force of Hume's argument, they did not accept his conclusion that science is an irrational pursuit and that therefore no general empirical knowledge is possible. That science was a rational affair and that it produced the only knowledge worth having was the basis on which their epistemology was constructed. They reconciled the scepticism of Hume with their own positivism by construing induction probabilistically. They gave up the classical empiricist demand that knowledge must be fully justified or proved with certainty. The adoption of physicalism as a theory of perceptual knowledge allowed for observation-language assertions to be checked by intersubjective agreement. If observers agreed about their findings then they had a high probability of being true. The demand for incorrigibility in knowledge claims was replaced by the demand for a high probability that a belief was true. The conclusion of an inductive inference was that it was probably true as a result of an accumulation of confirming evidence. Thus strict positivism gave way to a more moderate logical empiricism in which the verificationist theory of meaning was abandoned for a theory of confirmation. Carnap acknowledged the impossibility of conclusively verifying a scientific theory and replaced it with the notion of a 'gradually increasing confirmation' (Carnap, 1953, p. 48). This admission not only raises the question of the relationship between a scientific law and the observation statements against which it is tested, but it also resurrects the whole problem of how scientific terms acquire meaning. If meaning is no longer equated with empirical verification, then it appears to have an irreducible metaphysical element.

Another move to counter Hume's scepticism consisted in widening the concept of rationality to include inductive as well as deductive inference. Hume's contention that science was irrational was derived from his notion of rationality. Although he is depicted as the great exponent of empiricism there is a sense in which Hume had not succeeded in shaking off the influence of Aristotelianism in science. The deductive logic of the syllogism, which Bacon had rejected as a means of scientific discovery, was for Hume the paradigm of rationality, and it was by this standard that he rejected Baconian inductive science as irrational. Bacon, on the other hand, in moving from deductive logic to inductive inference, was setting a fresh standard of rationality. At one level what Hume had demonstrated was simply that the natural world is a contingent world, and that the relationship between objects and events in the natural world is not like that between symbols or statements in a deductive system. It is simply a confusion to demand that a factual proposition be an analytic truth. The world is just as it is, and it is a contingent affair. The law of the excluded middle ('if p' then 'p or not p' is true as a matter of logical necessity), which is central to deductive reasoning, is a fact necessary to our ability to reason. The law of the excluded middle and the contingency of the world are just given. If natural objects and events were related to each other in the same way as are the symbols in a closed deductive system, then the world would not be the world we know. The logical necessity that pertains in deductive argument does not pertain in the physical world. Inductive reasoning is as essential an ingredient of rationality as is deductive reasoning. Braithwaite, for example, writing in *British Philosophy in the Mid Century*, stoutly defends induction by maintaining that the logician has no need to give reasons for inductive inference (1966, p. 109). To question induction is like questioning deduction. This is precisely the stance taken by Ayer, who, because of the force of Hume's argument, turns from justification to description. Following Russell he uses the principle of induction to set the standard of rationality (Ayer, 1956, pp. 75-80). Referring to Hume, Russell admitted that 'Subsequent British philosophers rejected his scepticism without refuting it.' He concluded, 'Without the principle of induction science is impossible' (Russell, 1946, Ch. 17).

The admission of inductive logic as essential to the Logical Positivists' philosophy of science considerably weakened the authoritative base on which the Received View was constructed. The whole tenor of the positivist argument is to establish sure foundations on which knowledge can be built, but as Russell commented, 'Pure empiricism is not a sufficient basis for science' (Russell, 1946, p. 699). Although their philosophy of science remained justificatory, continuing to emphasise the importance of the confirmation of theories, the hint of fallibilism was present, and this finally proved to be fatal to the Received View as developed by the Logical Positivists.

The Received View arose out of the Logical Positivists' attempt to provide an explication of a scientific theory. They attempted to provide an exact and precise epistemology for natural science. This characterisation of science was an attempt to be both descriptive and prescriptive. It is descriptive in the sense that the positivists attempted to give an account of how theories are actually formulated, and it is prescriptive in that it excludes from science all theories that cannot be reformulated according to the canons of the Received View. The explication is based on a dogma: the dogma of empiricism which insists on reducing the scientific enterprise to observation and equating meaning with empirical verification. Empiricism fails at many points, as we shall argue, but the fundamental failure of empiricism is its inability to demonstrate why sense experience is necessary for scientific knowledge. It is an assumption essential to empiricism, but it is an assumption taken for granted and nowhere argued. This dogma of empiricism has important and far-reaching consequences for religion.

Metaphysics and Religion

It was the avowed aim of the positivists to eliminate metaphysics from science. The key to this endeavour was the conviction that observation or sense experience, alone, yielded authoritative knowledge. From Bacon's injunction to observe nature and make inductive generalisations, to Schlick's attempt to equate meaning with empirical verification, the effort was to construe science with no trace of metaphysical elements. We have seen how the implications of this doctrine went far beyond the confines of science and were spelled out in the Verification Principle. Here was a criterion that provided the demarcation between sense and nonsense, not just for science, but for all discourse. If metaphysical assertions were alien to the pursuit of scientific knowledge, then they were alien to the pursuit of knowledge in general, because scientific knowledge was the paradigm of all knowledge. According to the Verification Principle only scientific, that is empirically verifiable, assertions could count as knowledge.

In order to assess the status of traditional knowledge claims which conflicted with the Verification Principle, recourse was made to Kant's analytic-synthetic distinction. Various attempts have been made to construe this distinction, but the one the positivists favoured consisted in identifying analytic truths as truths of logic, and synthetic truths as the empirical truths of science. Thus the truths of mathematics and logic were tautological because they consisted in definitions and they were therefore necessarily true. According to Wittgenstein's dictum in the *Tractatus*, 'The propositions of logic are tautologies. The propositions of logic therefore say nothing. (They are analytical propositions.) . . . The propositions of mathematics are equations, and therefore pseudo propositions. Mathematical propositions express no thoughts' (6.1-6.21). Any non-tautological proposition that is in principle unverifiable by empirical means is, *ipso facto*, devoid of meaning. This rules out metaphysical statements in general and theological statements in particular. Such statements are neither true nor false but simply meaningless, and there is no point in attempting rational argument in such matters. 'It cannot be significantly asserted that there is a non-empirical world of values, or that men have immortal souls or that there is a transcendent God' (Ayer, 1946, p. 31). Ethical statements, according to this positivist dictum, are simply emotive utterances indicative of subjective attitudes, or simply slogans or exclamations intended to have a psychological effect on people. Moral discourse was construed as being nothing more than an appeal to people's emotions, and therefore irrational.

Metaphysics has been a central concern of Western philosophy from the Greeks onwards. The word 'metaphysics' was coined by students of Aristotle who used it to refer to his treatise on the subject he considered after (*meta*) physics. In this study, Aristotle was concerned with 'Being qua Being': that is, with the attempt to characterise existence or reality as a whole, rather than in its specific aspects considered in the various sciences. Kant was more concerned with epistemology than ontology; his metaphysics consisted in an attempt to establish the presuppositions on which all knowledge was based. Collingwood has more recently attempted an historical or sociological account of what in fact were the necessary characteristics of the presuppositions underlying the thought of particular groups or communities (Collingwood, 1940). Even more recently Strawson has distinguished between *Descriptive* and *Revisionary Metaphysics*; the former which is 'content to describe the actual structure of our thought about the world', and the latter which is concerned 'to produce a better structure' (Strawson, 1959, p. 9).

The Logical Positivists used the term 'metaphysics' to refer to all utterances which could not be construed as either analytic or synthetic in their vocabulary. They thus defined metaphysics, in the words of Ayer, as that which 'purports to express a genuine proposition, but does, in fact, express neither a tautology nor an empirical hypothesis' (Ayer, 1946, p. 41). The significant point here is that the positivists ruled out metaphysics as nonsense by definition and not by argument. And in so doing they rejected much traditional philosophy as a confused and unfruitful enterprise, confining philosophy to the logic of science (Ayer, 1946, p. 153; and Wittgenstein, 1951, 6.53).

Hume had anticipated this conclusion in a famous passage in his

Enquiry:

> When we run over libraries persuaded of these principles, what havoc must we make? If we take in our hand any volume; of divinity or school metaphysics, for instance; let us ask *Does it contain any abstract reasoning concerning quantity or number?* No. *Does it contain any experimental reasoning concerning matter of fact and existence?* No. Commit it then to the flames: for it can contain nothing but sophistry and illusion. (Hume, 1951, Section 12)

Thus only empirical science could provide knowledge. There could be no knowledge of a reality that transcended the world of science. And science itself was restricted by the canons of empirical verifiability. Quite clearly there could be no room for theology, which along with 'divinity and school metaphysics' was worthy only of the flames.

Ayer spelled out the implications of this thesis in his *Language, Truth and Logic* (1946, pp. 114-20). He argues that the term 'God' is a metaphysical term. And if 'God' is a metaphysical term then it cannot be even probable that God exists, for to say that 'God exists' is to make a metaphysical utterance which cannot be either true or false. And by the same criterion no sentence which purports to describe the nature of a transcendent God can possess any literal significance. Ayer does not wish his position to be confused with that of the atheist or agnostic because in joining issue on the question of God's existence they allow meaning or significance to propositions about God, whereas Ayer denies that 'God talk' has any significance whatsoever. For the atheist and agnostic the question of God's existence is a genuine question; for Ayer it is meaningless. God is not a genuine name. 'There cannot be any transcendent truths of religion. For sentences which the theist uses to express such "truths" are not literally significant.' God-talk is nonsense-talk. Unless the theist can 'formulate his "knowledge" in propositions that are empirically verifiable, we may be sure he is deceiving himself'. People who speak as if they 'know' moral or religious truths are 'merely providing material for the psychoanalyst'. 'For no act of intuition can be said to reveal a truth about any matter of fact unless it issues in verifiable propositions. And all such propositions are to be incorporated in the system of empirical propositions which constitutes science.'

Although Ayer has long since departed from the arguments employed in *Language, Truth and Logic*, he still maintains that 'the approach of the book was right' (Ayer, 1971, p. 55),[10] and his refined brand of empiricism allows little, if any, room for religion. His more recent thoughts on the subject are contained in *The Central Questions of Philosophy* (1976, Ch. 10). His discussion of the claims of theology centre on the question of the existence of God. This is a favourite starting point for sceptical philosophers who feel it necessary to establish the reality and rationality of God's existence before they can proceed to entertain and discuss the rationality of religious beliefs and practices. This requirement on the part of the empiricist should not surprise us. Empiricism is, after all, based on the reality of sense experience from which one infers the existence of material objects and events about which the scientist can form theories and with which he can conduct experiments. Admittedly Mach's extreme sensationalism offers only a psychological account of science, but subsequent empiricists are realists in the sense that they affirm the existence of physical objects and events; they affirm the reality of the physical world which is the object of scientific enquiry. They are strongly influenced by the outlook of physicists themselves who assume the reality and rationality of the external world (Ayer, 1976, p. 110). But the empiricist fails to demonstrate the reality and rationality of the external world. Ultimately he opts for the common-sense view on which natural science has grown. It boils down to a *decision* he makes. And yet the empiricist gets nowhere in theological discussion because he is unable to account satisfactorily for the existence and rationality of belief in God. The religious person could follow the example of the empiricist and say that he is influenced by the outlook of religious people who assume the reality of God. The theist, like the empirical realist, makes a decision, and on that decision he bases his theology and his practice.

The logical empiricist never allows the argument for theism to get started. If, as he stipulates, the only thing that can count as knowledge is that which is in principle empirically verifiable, then knowledge of a non-sensory object, that is God, is ruled out by definition. Knowledge of God does not count as knowledge because it does not figure in the empiricist's outlook. On the one hand the empiricist insists that the religious enterprise must start with establishing the existence of the object of theological enquiry, that is God; and on the other, the existence of God is ruled out on the grounds that it is not an empirical proposition. In other words, theology is dismissed because it is not science. But of course, if it were science, it would not be theology. Quite clearly, if the positivist account of science and the growth of knowledge, based on observation, inductive inference and empirical verification, is correct, then religion has had its day. It is simply not part of empirical science, and therefore no claimant to knowledge. But if religious talk were scientifically verifiable, after the logical empiricist tradition, it would no longer *be* religious talk, but part of natural science.

The Failure of the Verification Principle

It was largely through the influence of the Verification Principle that the positivist analysis of science was extended beyond the realm of science to general knowledge claims. Like the Received View as a theory of science, the Verification Principle had a wide and lasting influence. However, as a principle of philosophy it had a short, if startling, life. It would be tedious to rehearse in detail all the arguments that have been marshalled against the Verification Principle, especially as the principle underwent progressive modification in the face of difficulties raised by its critics. Isaiah Berlin pointed out as long ago as 1938 that 'the principle of verifiability or verification after playing a decisive role in the history of modern philosophy . . . cannot . . . be accepted as a final criterion of empirical significance, since such acceptance leads to wholly untenable consequences' (Berlin, 1968, p. 15). His article 'Verification' remains one of the classical criticisms on the subject. And yet, as we shall later indicate, philosophers and theologians a quarter of a century later were still trying to adapt their theology to his presuppositions.

A preliminary skirmish with the principle exposes it to the charge of circularity. Before one can begin to verify a given statement one must first know what it means, otherwise one could not set about the process of verification. Therefore meaning is in some sense prior to verification and not simply equivalent to it.

A more serious objection was developed by Isaiah Berlin when he established that meaningless statements could be made to conform with the principle. Thus absurdities such as 'The Absolute is lazy', or 'This logical problem is bright green' are allowed meaning by Ayer's definition of verifiability (Berlin, 1968, p. 22). Further attempts to amend the criterion so as to escape ingenious logical criticism, for instance by Ayer in his 1946 introduction to *Language, Truth and Logic*, were all unsuccessful.[11]

Statements about the past, such as 'Caesar crossed the Rubicon,' also presented insurmountable difficulties. According to the Verification Principle statements about the past (or future) had to be translated into statements about present experiences. But the two sets of statements can never be equivalent.

In his first formulation of the Verification Principle Ayer had recognised that because of the problem of induction no general principle could be established with certainty and so he distinguished between what he called strong and weak verification. Strong verification demanded conclusive evidence from experience; but this could only result from a single, incorrigible sense experience and was therefore valid only for the experiencer. In order to save the general propositions of science from being assigned to the metaphysical rubbish heap (because they could not be conclusively verified) Ayer had to be satisfied with weak or inconclusive verification, thus sacrificing certainty for probability. But although he made this distinction between strong and weak verification, in practice he confused the two because even weak verification depends upon incorrigible, individual experiential propositions, that is on strong verification. In other words, even in its weak sense, the Verification Principle could not function neutrally, but depended upon the subjective experiences of the observer. Ayer now admits that

> we failed to make it clear, even to ourselves, whether 'the method of verification' had to lie within the possible resources of the person who was interpreting the proposition . . . or whether the requirement of verifiability could be construed in a more impersonal and therefore more liberal fashion.
>
> In the first edition of *Language, Truth and Logic* I put forward the Verification Principle only in its weaker aspect, as supplying a means of demarcating sense from nonsense, but actually used it in the stronger way when I attempted to give analyses of various types of proposition. In both cases, I adopted the standpoint which I think that Ryle was the first to criticise as that of 'Verifiability by me'. (Ayer, 1979, p. 325)

Hence the move to verifiability in general from verifiability by me does not simply weaken the Verification Principle but renders it incomprehensible because verifiability in general depends upon verifiability by me. The point here is that verifiability in general involves the attribution of experiences to others and this in turn depends upon analysis of their overt behaviour. Carnap opted for this physicalist or behaviourist thesis which, as we have noted, became incorporated into the Received View of scientific theories. Phenomenalists such as Ayer tried to combine this behaviourist analysis of the experience of others with a mentalistic account of one's own experiences which, as Ayer admits, is an unacceptable inconsistency (Ayer, 1979, p. 326). The positivist who adopts the phenomenalist thesis must retire to his corner muttering only to himself: a far cry from the impersonal and liberal tradition of science which the positivists set out to establish.

Bhaskar, in his *A Realist Theory of Science*, maintains that the Logical Positivists committed what he calls the *epistemic fallacy* in formulating the Verification Principle. The epistemic fallacy is a metaphysical dogma which confuses statements about being with statements about our knowledge of being. If we are to maintain a realist view of science then we must not reduce ontological questions to epistemological terms.

> Verificationism indeed may be regarded as a particular form of the epistemic fallacy, in which the meaning of a proposition about reality (which cannot be designated 'empirical') is confused with our grounds, which may or may not be empirical, for holding it.
> (Bhaskar, 1978, p. 37)

The ultimate irony concerning the Verification Principle is that as a principle of demarcation between sense and nonsense (a criterion of meaning) it is a highly metaphysical formulation. Wittgenstein recognised the self-refuting nature of the propositions of his *Tractatus* when he wrote at the end:

> My propositions are elucidatory in this way: he who understands me finally recognises them as senseless, when he has climbed out through them, on them, over them. (He must so to speak throw away the ladder, after he has climbed up on it.) (6.54)

Ayer has since admitted puzzlement that he and his fellow positivists did not notice at the time that the Verification Principle contradicts its own assumptions, being itself neither a tautology nor an empirical assertion. In Germain Grisez' pungent phrase, it 'self-destructs' (quoted by Meynell, 1977, p. 26). Ayer's attempt to recover from this extraordinary blunder consisted in offering the Verification Principle as a 'stipulative definition', but as he later recognised, one is under no obligation to accept the stipulation (Ayer, 1971, p. 56). The underlying problem with all empiricist attempts to reduce knowledge to observation statements is that even the statement of such a principle is not allowed by the principle itself.

The question of the status of the Verification Principle, if it is neither analytic nor synthetic, raises the whole question of how to classify philosophical propositions. Ayer suggested that much of philosophy consists of persuasive definitions, or that philosophical statements are *sui generis* and as such explanatory or clarificatory. His admission that 'the notions of what it actually is to clarify a concept or explain a theory or to justify a belief are obscure' (1971, p. 57) is a far cry from the original positivist claim for certainty based on empirical knowledge.

This obscurity is not clarified by Ayer's attempt to distinguish between *primary systems*, containing purely factual propositions describing what is actually observable, and *secondary systems*, which contain terms not directly related to anything observable but which have explanatory power. Here is an attempt to distinguish an area of meaningful metaphysics, and in so doing Stephen Körner comments, 'Ayer is no longer fighting a rearguard action for logical positivism but attacking one of its central positions' (Körner, 1979, p. 267). Hence metaphysics is now allowed as a 'secondary system', and the term 'metaphysical' no longer has a pejorative force. The reformed Ayer even allows 'the possibility that the positing of a superior intelligence as the author of nature should serve as an explanatory hypothesis', although he hastens to add that it is not one which he is prepared to accept (Ayer, 1979, p. 331).

The Demise of the Received View

In outlining the development of the Received View we noted how the original strict positivistic programme was allowed a more liberal interpretation, but further amendments proved finally to be fatal to this theory of science.

A fundamental problem for proponents of the Received View was that it immediately excluded from genuine science a great number of theories which were generally referred to as scientific. Suppe lists as examples:

Hebb's theory of the central nervous system, Darwin's theory of evolution, Hoyle's theory on the origin of the universe, Pike's tagmemic theory of language structure, Freud's psychology, Heyerdahl's theory about the origin of life on Easter Island, or the theory that all Indo-European languages have a common ancestor language, proto-Indo-European . . . Most theories in cultural anthropology; most sociological theories about the family; theories about the origin of the American Indian; most theories in paleontology; theories of phylogenetic descent; most theories in histology, cellular and microbiology, and comparative anatomy; natural history theories about the decline of the dinosaur and other prehistoric animals; and theories about the higher processes in psychology. (Suppe, 1977, p. 65)

The reason for this high casualty rate amongst 'scientific' theories resulting from the adoption of the Received View is that such theories do not admit of axiomatisation, or simple deductive structure, which is essential to the Received View's analysis. Such theories are insufficiently precise or developed to permit the reduction to a system of interrelated basic concepts and axioms and definitions (or correspondence rules). This is not surprising when we recall that the Received View arose out of an examination of the theories of physics. But if only the theories of physics are counted as scientific, and as such the only true claimants to reliable knowledge, then large areas of human intellectual endeavour must be consigned to the metaphysical rubbish heap. Clearly this will not do.

Even if the reductionist programme is successfully carried out and, as Carnap envisaged, 'everything can be reduced to physics' the question still arises as to the adequacy of the Received View as an analysis of physical theories.[12]

We have already indicated the modifications made to the Received View which were consequent upon criticism by its proponents whose intention was to refine and improve the analysis. As this criticism developed it became apparent that the Received View was fundamentally untenable and it finally gave way to alternative attempts to characterise the scientific enterprise. The force of these criticisms will become more evident when we consider some of the alternative analyses which arose in conjunction with the demise of the Received View.

The positivist programme for science ultimately depends upon a clear distinction between observational and theoretical components of a theory. Although, as we have observed, theoretical or non-empirical components were admitted in scientific theories, the main burden of the positivist analysis rested on reducing the components of scientific theories to observational statements. Such statements were construed as synthetic because factually significant, whilst the logical components of the theory were construed as analytic. Hence the observational-theoretical distinction rested in turn on the analytic-synthetic analysis. Central to the positivist epistemology is the thesis that every cognitively significant or meaningful sentence is either analytic or synthetic. One of the first to expose the naivety of this distinction was Popper (1934), but the debate continued well into the 1960s with the work of Quine,

Hanson and Putnam.

Quine, in his 'Two Dogmas of Empiricism' (1953), maintained that the analytic-synthetic distinction is untenable, and Putnam introduced the idea of law-cluster concepts which were neither analytic nor synthetic, as essential to scientific theories (see Suppe, 1977, pp. 73-6). Hanson (1958) developed the early Popperian thesis that all observation is theory impregnated. This considerably weakened Carnap's characterisation of synthetic sentences as those whose truth or falsity can be directly determined by factual information about the world. If such factual information is not clearly and readily available as conclusive evidence for or against a theory then the empiricist methodology is seriously undermined. The proponents of the Received View allowed a theoretical component in their analysis of scientific theories, but once it is shown that the factual or empirical statements themselves have a theoretical component, then the positivist programme collapses. Observation is not such a straightforward, cognitively significant exercise as had been supposed.

Once the observational-theoretical distinction has been blurred, the notion of partial interpretation of theories becomes redundant. The amendment of the Received View by its proponents to admit partial interpretation by correspondence rules was allowed in order to accommodate an irreducible theoretical element in a scientific theory. If a clear distinction cannot be maintained between the observational and theoretical terms of a theory, then even the partial reduction of theoretical terms to observation terms makes little sense. This raises the whole question of the function and status of the correspondence rules, or implicit definitions, which are central to the positivist analysis of the Received View. The role of the correspondence rules was to define and thus guarantee the cognitive significance of theoretical terms and to specify the experimental procedures for applying the theory to particular phenomena. The difficulty of maintaining the observational-theoretical distinction eliminated the first of these functions, thus putting all the weight on the specification of experimental procedures. But this is not such a simple and straightforward matter as it first appeared to proponents of the Received View. Correspondence rules are not simple definitions contained within the bounds of a single theory, but more like auxiliary hypotheses which themselves depend on further theoretical assumptions. Any statement contained within a scientific theory is a part of an indefinite causal chain of dependent statements. Thus the role of correspondence rules in the Received View is clearly unsatisfactory, and the reductionist programme, dear to the hearts of its proponents, cannot be carried out.

The positivistic ideal contained in the Received View is that a theory should have a simple deductive structure, the ideal form of which is exhibited in mathematical thinking. The cognitive significance, or meaning, of the terms of a theory depends on the deductive links contained within the system. It is the failure to maintain these deductive links and to retain a simple deductive structure that exposes the Received View as an inadequate and misleading account of the structure of scientific theories. This problem of reduction, when coupled with the difficulties already alluded to concerning explanation, induction, verification and confirmation, ultimately led to the abandonment of the Received View.

Suppe, after a detailed critical analysis of the Received View, maintains that 'the last vestiges of positivistic philosophy of science are disappearing from the philosophical landscape' (Suppe, 1977, p. 619). He concludes that

> virtually all the positivistic programme for philosophy of science has been repudiated by contemporary philosophy of science. The Received View has been rejected, as have its treatments of explanation and reduction . . . Positivism today truly belongs to the history of the philosophy of science, and its influence is that of a movement historically important in shaping the landscape of a much-changed philosophy of science. (Suppe, 1977, p. 632)

The Received View was based on a dogma. It was an attempt to provide an analysis of how theories ideally should be formulated. It proved to be an ideal that theories in practice failed to meet, because scientific theories are much more complex than the positivist programme envisaged, or could by its principles allow. It was an attempt at a rational reconstruction of science: more accurately it was an attempt both to identify and specify *the* rationale of science. Hence the Received View, presented as *the* theory of science, implied that the 'scientific method' (as conceived by the Received View) is the paradigm of rationality, with the implication that all that is not science is irrational. Consequently 'scientific' became the synonym for 'rational' or 'reasonable'.

Although the positivistic programme for the philosophy of science has been repudiated by contemporary philosophers of science, it remains the unquestioned view of science held by many practising scientists and laymen. As Suppe observes, 'It seems to be characteristic, but unfortunate, of science to continue holding philosophical positions long after they are discredited' (Suppe, 1977, p. 19). Putnam also comments that 'Scientists tend to know the philosophy of science of fifty years ago' (Putnam, 1978, p. 235).

An example of such an attitude is contained in the University of North Wales, Bangor, *Prospectus* for 1981-2 in which it is stated that science is 'the accumulation and organisation of knowledge acquired by observation'. Such a statement takes us right back to Bacon himself, with whom we began our enquiry, and it takes no account of the great movement of ideas which has since taken place. If contemporary practising scientists can be so naive in their view of the theory of science, it is not surprising that contemporary theologians make the same erroneous assumptions.

The Theological Response

Despite the fact that the Logical Positivist thesis ultimately proved abortive, the Verification Principle abandoned, and the Received View totally discredited, such has been the force and appeal of positivism and its heir, logical empiricism, that it continued to set the standard of rationality long after its demise. The appeal of positivism is the seductive appeal of authority, luring man to the tree of knowledge and ridding him of doubt and uncertainty. Its presumptions have had a deep and lasting influence on the general intellectual climate of our time. In particular, modern scepticism towards religion is founded very much upon positivistic assumptions. The general assumptions of empiricism, in particular its notions of empirical evidence, justification and verification as pre-requisites for any knowledge claims, were inherited by a generation of theologians and philosophers who valiantly tried to square the claims of religion with the presumptions of positivism. The rise of Modernism and Liberal Protestant Theology in the late nineteenth and early twentieth centuries was in no small measure a result of the anti-metaphysical intellectual climate created by positivism in science and philosophy. Because science was a success story in terms of its theoretical achievements and practical applications it was uncritically assumed by Biblical critics and theologians that the theory of science, that is its philosophical foundation, was beyond question. It became almost sacrilegious not to defer reverentially to the authoritative claims and methods of science. That there was a 'scientific method' and that this method provided the only means of achieving reliable knowledge was an assumption readily made, and it invested the researches of many scholars with credibility and respectability. Not least did it influence the methods of historical criticism, which in its turn became 'scientific' and positivistic.

Consequently, liberal theologians attempted to subject the New Testament writings, which had formed the basis of Christian doctrine, to more rigorous scientific methods of criticism. Harnack, Schweitzer and Bultmann, among others, looked afresh at the New Testament, and the end result of this examination was an attempt to demythologise these foundation documents of the Christian faith, and to reinterpret them without the supernatural element.[13]

The theologian Karl Barth would have none of this, and there is a sense in which he played the positivists at their own game by adopting a positivist stance in his theology. Barth considered natural theology to be a futile exercise and he maintained that man cannot establish anything about the existence and nature of God from rational argument. Revelation alone is supreme. The concrete sources of this divine revelation are the Word of God in scripture and in the incarnate Son of God.

Barth was not concerned with the intellectual defence of a metaphysical system which would support religion. Instead of looking for a philosophical justification of the metaphysical presuppositions of religion, Barth, like the positivists, directs our attention to the given, that is to the concrete, historical account of God's revelation in the Bible, and particularly in the person of Christ. He thus accomplishes the liberation of Christianity from metaphysical propositions by the application of what Donald MacKinnon has called 'Christological Positivism'. Men are directly confronted with the revelation of God in Christ, and that calls for response. Barth's rejection of the metaphysics of religion, and of a natural theology that attempts to arrive at a knowledge of God though the world, is a refusal to take issue with the positivists on their own ground. But like the positivists he is always the champion of the concrete against the abstract or merely possible. Just as 'the law of causality' was central to classical physics, so is the person of Christ central to Christian theology. Barth exalts Christology above theology and philosophy. MacKinnon comments:

> There is certainly a note of positivism here, something analogous to that sounded by Bertrand Russell when he said, suggesting the application of the methods of mathematical logic to the solution of classical philosophical problems: 'whenever possible, let us substitute logical constructions out of the observable for inferred, unobserved entities'. There is in Barth something analogous to this recommended logical economy. We must, he insists, substitute for abstract, general statements concerning the being and purposes of God, and of

men, statements that show them in terms of, or set them in relation to, Jesus Christ. (MacKinnon, 1968, p. 68)

Barth's repudiation of the metaphysical element in religion thus shares much in common with the Logical Positivists' repudiation of the metaphysical element in science. The dogmatic approach is common to both. For the Logical Positivist the standard of rationality is set by science; for Barth it is irrelevant. There can be no rational appraisal of God's revelation, it is the given, one either accepts it or not. This revelation of God, in his Word, enters the world in judgement upon all things human; not least on man's reasoning about God. Barth wastes no time in arguing with the positivists about metaphysics and meaning, and he wastes no time in arguing about God. His task is to interpret God's revelation in Christ in terms of the affairs of men and their relationship with God.

Thus Barth remained unmoved by the positivist challenge, but the final outcome of the religious subjectivity of liberal theology was a religion without propositions, an emphasis on personal 'existential' faith, and the exaltation of love above all theological and doctrinal utterances.[14] The Church was thus deprived of much of its dogmatic and ecclesiastical authority. As David Edwards comments:

such liberal Protestantism has remained the simple creed of many millions of Protestant laymen. Many of these have, indeed, maintained that to be a Christian it is not necessary 'to go to Church' at all . . . the Church has been understood as a fellowship of laymen trying to live decent lives with help from stories about Jesus and from such worship as has appealed to individual temperaments or congregations. (Edwards, 1969, p. 168)

Alasdair MacIntyre was even more scathing when he wrote:

The difficulty lies in the combination of atheism in the practice of the life of the vast majority, with the profession of either superstition or theism by that same majority. The creed of the English is that there is no God and that it is wise to pray to him from time to time. (MacIntyre, 1963, pp. 227-8)

Coupled with the rejection of the supernatural in religion was the rejection of attempts to explain life in the universe as a whole as an excess of metaphysical speculation. Such 'ultimate' questions had been ruled out of science by positivism, and this prohibition was now extended to religion. John Wisdom, in his essay on 'Gods' (1953), introduced the parable of the gardener which was further developed and exploited by A.G.N. Flew (1971).

In Wisdom's parable two people return to their long-neglected garden and find that it is still in good order. Consequently one affirms and the other denies the existence of a gardener who tends and cares for the garden. Wisdom uses this parable to indicate the nature of religious commitment and to illustrate the impossibility of establishing one way or another the existence of God. Flew's extension of the parable leads him to reject as absurd the assertions of religious believers about God's existence and activity. No material evidence can support the believer's claims. 'Just how does what you call an invisible, intangible, eternally elusive gardener differ from no gardener at all?' He concludes that religion 'dies the death of a thousand qualifications' (Flew, 1971, pp. 13-14).

Flew's scepticism regarding religion is not far removed from that of Ayer, and both find their origins in the positivistic programme for science. Flew replaced Ayer's demand for verification with that of falsification, and he rigorously develops this thesis in a later book, *God and Philosophy* (1966). Unlike Ayer, who originally refused to argue with the religious believer on the grounds that he was talking nonsense, Flew accepts the challenge. His aim is to destroy the credibility of orthodox doctrinal theism by rational argument; albeit a rationality defined by a theory of science. His claim is that religious assertions are incoherent because they allow of no possible refutation. Such refutation, or falsification, must be grounded in empirical evidence. The programme for science becomes the progamme for religion. The authority previously ascribed to religious tradition is rejected in favour of the authority of the new tradition of science based on the authority of the senses. The mid-twentieth-century attempts to secularise theology were the last-ditch attempt by theologians who had too readily assumed this transfer of authority and with it the definition of rationality supplied by positivism and its successor, logical empiricism.

One such attempt was made by the philosopher of science, R.B. Braithwaite, in his 1955 Eddington Memorial Lecture, 'An Empiricist's View of the Nature of Religious Belief' (1971, pp. 72-91). Braithwaite himself was an exponent of the Received View, a version of which he presented in his *Scientific Explanation* (1953, Ch. 2).[15] He opens his 1955 lecture by quoting Eddington: 'The meaning of a scientific statement is to be ascertained by reference to the steps which would be taken to verify it.' This verificationist stance, Braithwaite maintains, 'is in complete accord with contemporary philosophy of science'. Braithwaite makes it clear that he accepts the interdependence of meaning and verification not only for science, but also for religion. Consequently the question of meaning in theological statements is logically prior to the question of truth. 'A religious statement cannot be believed without being understood, and it can only be understood by an understanding of the circumstances which would verify or falsify it.' If religious assertions are neither analytic (that is necessarily true) nor synthetic (that is verifiable by standard methods), 'Does this imply that religious statements are not verifiable, with the corollary, according to the Verification Principle, that they have no meaning and, though they purport to say something, are in fact nonsensical sentences?' Braithwaite points out that moral statements, as well as religious statements, are unverifiable by standard methods, but they are clearly meaningful as guides to conduct. He then makes the Wittgensteinian substitution of *use* for verification, 'the meaning of a statement is given by the way it is used', and suggests that religious statements are really declarations of moral policy in the guise of factual and historical statements. Braithwaite thus invests religious belief with meaning by reducing it to morality. If we are allowed ethics without propositions, why not religion without propositions?

> I myself take the typical meaning of the body of Christian assertions as being given by their proclaiming intentions to follow an agapeistic way of life, and for a description of this way of life – a description in general and metaphorical terms, but an empirical description nevertheless – I should quote most of the thirteenth chapter of I Corinthians.

If two or more religions can be reduced to similar moralities, then they are to be distinguished by the different sets of stories which are associated with them. The stories are rather like Bunyan's *Pilgrim's Progress* or the novels of Dostoevsky: they are literally untrue, but they do influence our behaviour. Braithwaite concludes that morality 'is of the very essence of the Christian religion'. His thesis is not far removed from that of the American philospher John Dewey, who twenty years earlier interpreted religion without the supernatural as a code of moral conduct associated with religious customs (Dewey, 1934).

However, this reduction will not satisfy the religious man who quite plainly does think that his beliefs have something to do with his relationship with God and with ultimate questions regarding his place in the universe. Braithwaite has been forced into this position by his whole-hearted acceptance of positivism both in science and in its extension in general epistemology.

The theologian Paul van Buren makes a similar, but more detailed, reduction of religion from talk about the supernatural to talk about man in his book *The Secular Meaning of the Gospel* (1963). He develops a secularised theology which attempts to reinterpret the tradi-tional assertions of Christianity in such a way that they become asser-tions about 'this world'. Van Buren sets forth a version of Christianity in which there is no such being as God and in which the miraculous event of the resurrection of Jesus Christ has no place.

Van Buren's secular man is the child of positivism. He has inherited a scepticism which finds no place for metaphysical ideas or any trace of the supernatural or miraculous. At the same time he has inherited the rich tradition of the Christian faith. Van Buren attempts a reconcilia-tion between these two uncomfortable bed-fellows by suggesting that the believer's assertions about God and sin and prayer and grace are really assertions about human freedom and love.

Van Buren takes up a challenge from Bonhoeffer in his effort to make sense of religion for contemporary man who has 'come of age'. He sides with the sceptic in Flew's parable of the gardener in rejecting God-talk as meaningless. The assertion that God-talk is meaningless is of course based on the presumptions of positivism which van Buren at no point challenges. He unquestioningly accepts that 'the whole tenor of thought of our world today makes the biblical and classical formula-tions of [the] Gospel unintelligible' (p. 6). Van Buren admits that he 'has taken certain empirical attitudes characteristic of modern thought seriously and accepted them without qualification', and this brings us face to face with 'the difficulty of finding any meaningful way to speak about God'. He naively accepts that 'we share certain of the empirical attitudes reflected in the "revolution" in modern philosophy', and these empirical attitudes constitute the 'secular man' (p. 156).

The key to the resolution of van Buren's dilemma is 'linguistic analysis' which will reveal 'the meaning of the Gospel' (p. 17). Van Buren clings to this heaven-sent aid as the life-raft that will keep Christ-ian belief afloat. Linguistic analysis is an offspring of positivism and it makes the same authoritative claims to soundly based knowledge. Its use reveals the true meanings of words and assertions. Van Buren accepts Flew's dictum that 'the meaning of [a word] can be elucidated by looking at simple paradigm cases: such as those in which fastidious language users employ [that word] when the madness of metaphysics is not upon them' (p. 71).[16] 'The madness of metaphysics' is manifested by the application of the positivist Verification Criterion of Meaning which van Buren adopts in a modified form:

> the meaning of a word is its use in its context . . . If a statement has a function, so that it may in principle be verified or falsified, the statement is meaningful, and unless or until a theological statement can be submitted in some way to verification, it cannot be said to have a meaning in our language game. (pp. 104-5)

Consequently the word 'God' is avoided because it is not a proper name (i.e. 'God' statements cannot be verified) (p. 145).

The only way in which theological statements can be submitted to verification is by revealing their true meaning. This is achieved when 'God statements' have been translated into 'man statements', and reference to God's activities interpreted as statements about human existence (p. 103). Hence 'the meaning of the Gospel is to be found in the historical and the ethical, not in the metaphysical or the religious' (p. 197).

Van Buren's hand is fully revealed when he applies this new-found method of linguistic analysis to the resurrection, which is central to the Christian faith, and concludes that it is an event in the life of the disciples who were made free. Easter 'was the story of the free man who had set them free', and the good news of the Gospel was the proclamation of this newly discovered freedom. The 'event of Easter' consisted in the death of Jesus on the cross and the subsequent realisation by some of his disciples of the meaning of his life and the liberating effect this had on them (p. 134).

Van Buren unwittingly accepts the reductionist thesis, central to positivism, that all meaningful language can be reduced to language about physical things. Just as astrology has been reduced to astronomy; alchemy to chemistry; so theology, by 'the rigorous application of the empirical method', is reduced to talk about man and his freedom (p. 198). The fact that talk about man and his freedom involves some highly metaphysical theses escapes van Buren's notice.

Van Buren's attempt to make theology respectable by reinterpreting it along empirical lines is an attempt to square the circle, and it inevitably leads to atheism. The transcendent God of classical Christianity has no place in his secular Gospel. Christian thinkers down the ages simply did not know what they were talking about. Van Buren, with the aid of linguistic analysis, has shown us what they really meant, although they did not realise it.

We have taken van Buren's book as an example of the effect of positivism in science and philosophy on theological thinking. Another American theologian, Thomas Altizer, taking up Nietzsche's 'death of God', made a similar attempt at producing a thoroughly secular and empirically grounded theology in *The Gospel of Christian Atheism* (1967). In Britain, John Robinson contributed to the debate in his *Honest to God* (1963), which was philosophically naive and which unwittingly accepted the empiricist account of science.[17]

The efforts were confused and the results absurd. The confusion and the absurdity arose out of the uncritical acceptance of positivism in science and its application as a general principle of philosophy.

Central to the positivist theory of science is the notion of verification. We have seen that the verification principle as an extension of this thesis became an arbiter between sense and nonsense and therefore an essential criterion of rationality. Theological assertions, because unverifiable, are thus rendered nonsensical (or according to Flew, vacuous because unfalsifiable), and therefore irrational.

Consequently, theologians who never questioned the soundness of this doctrine in science, and its extension as a general principle of rationality, felt uneasy about the problem of verification in religion. After all, religion involves a search for, and an encounter with, the Truth; but if assertions about God and his relationship with man cannot be verified, or even rendered highly probable by induction, then they are fit only for the Humean flames.

John Hick considers this problem in his article 'Theology and Verification' (1971, pp. 53-71). He admits that the theist cannot verify his claim about God's existence but makes a case for what he calls 'eschatological verification'. Verification, for Hick, consists 'in the ascertaining of truth by the removal of grounds for rational doubt'. That this involves empirical verification is evident when he enlists Schlick's support in claiming that survival after death is not a metaphysical problem but an empirical hypothesis. He has clearly not heeded Wittgenstein's aphorism in the *Tractatus* that 'Death is not an event in life' (6.4311).

Having argued that the Christian doctrine of the resurrection is not a self-contradictory notion, and that survival after death is subject to future verification, he then attempts to relate survival experiences with theism. Hick agrees that there are difficulties in any claim to have encountered God, but maintains that there could be after death experiences which would convince the believer. God's existence 'could be so fully confirmed in post-mortem experience as to leave no grounds for rational doubt as to the validity of that faith'. He concludes that 'the existence or non-existence of the God of the New Testament is a matter of fact, and claims, as such, eventual experiential verification'.

Hick's argument is distinctly odd. He is concerned to maintain the rationality of Christian theism. But his positivistic conception of rationality involves empirical verification. Such verification of the theist's claims being unobtainable in this world and in this life, Hick transfers it to the next. But whether or not the theist will, after death, find proof of his earthly beliefs, has nothing to do with the rationality of religion now, and this is the point at issue in the positivist's rejection of metaphysics and religion.

The implication of Hick's argument is that theism is irrational, because unverifiable now, but that in a future life the theist could be vindicated. His confusion arises out of an uncritical acceptance of verification as a criterion of rationality.

The recent revival of the argument about the Christian doctrine of the incarnation re-echoes to a large extent the efforts of the earlier liberal Protestant theologians to construct a non-metaphysical Christology based solely on the works and words of the 'historical Jesus'. *The Myth of God Incarnate* (Hick, 1977a), for example, provides us with a contemporary example of the influence of positivism on theological debate.

The authors of this collection of essays do not directly allude to a positivistic conception of science, but such assumptions are implicit in their attempt to dismiss incarnational language as unintelligible and nonsensical. Maurice Wiles, in the opening paper, 'Christianity without Incarnation?', asks whether incarnation is essential to Christianity. He questions the intelligibility of the doctrine and argues for its abandonment as a metaphysical claim about the person of Jesus. His anxiety to rid Christology of metaphysics is reminiscent of the positivists' efforts to rid science of metaphysics. There is here a strong note of naturalism which prevents him from taking seriously any notion of divine action in the world. His talk about God is such as to remove him entirely from the human scene. Don Cupitt, in his article 'The Christ of Christendom', and also in a subsequent book *Jesus and the Gospel of God* (1979), dismisses Trinitarian doctrine as a misguided metaphysical construct which was devised for social and political reasons. He too rejects any attempt to link Jesus philosophically or metaphysically with God. Like the Logical Positivists before him, Cupitt relegates all the metaphysical shibboleths of Christology to the dustbin. Cupitt's historical account of the development of the doctrine that became enshrined in the Chalcedonian Formula is highly suspect, but it is undergirded by a concept of rationality that went out with positivism. Cupitt's uncritical acceptance of the positivist view of science and the outmoded epistemology associated with it finally leads him to take leave of God, and to dismiss traditional spirituality as incoherent (Cupitt, 1980).[18] Further evidence of the positivist influence is evident in Denis Nineham's notion of history in his 'Epilogue' (1977). His scepticism feeds on the 'post-Enlightenment' sense of history in which the only historical statements allowed are those that are plainly descriptive or factual. This is a somewhat naive view of history which takes us back to early formulations of the Verification Principle which insisted that historical statements must be translated into statements about present experiences. Such a positivist view of history allows of no interpretation but only simple, basic, verifiable statements of fact. Nineham quotes with approval Dr Norman Perrin, who distinguishes between what he calls 'faith knowledge' and 'historical knowledge'.

> That Jesus died nobly or showed confidence in God are historical statements, subject to the vicissitudes of historical research, but that his death fulfilled the purpose of God in regard to 'my sins' is certainly not such a statement, and it lies beyond the power of the historian, even to consider it, even though as a Christian, he might believe it. (Nineham, 1977, pp. 197-8)

The problem for those who adopt such a 'post-Enlightenment' sense of history is that history becomes an impoverished, shrunken husk of what the great historians from Herodotus and Thucydides to Gibbon and Toynbee have made it. If only statements such as 'Caesar crossed the Rubicon,' 'Christ was crucified,' 'Napoleon was defeated at Waterloo' are allowed then the history of civilisation would be reduced to a single short volume, devoid of human interest and significance.

However, the fact that Perrin and Nineham allow that 'Jesus died nobly or showed confidence in God' as historical statements is evidence enough to demonstrate that this positivistic programme cannot be carried out. Whether or not Christ died nobly is as much a matter of speculation as is the question whether or not the hunger striker Bobby Sands died nobly. Some would say that both were foolish and deluded. The point is that if we allow the latter interpretive statements about the death of Christ or of Sands as historical, then we must allow statements such as those of Saint Paul, who wrote, 'God was in Christ, reconciling the world unto himself' (2 Corinthians 5.19. AV). The latter quotation from Saint Paul is of course highly theological, but historical research is not irrelevant to our assessment of its truth or falsity.

Hick makes precisely the same positivistic assumption when he writes:

> Is the statement that Jesus was God incarnate, or the Son of God, or God the Son, a statement of literal fact: and if so, precisely what is the fact? Or is it a poetic, or symbolic, or mythological statement? It can, I think, only be the latter. It can hardly be a literal factual statement, since after nearly 2000 years of Christian reflection no factual content has been discerned in it. (Hick, 1980, p. 55)

The implication here is that painstaking experiment over a period of time is required to establish whether or not a statement is factual. Hick's papers are sprinkled with terminology reminiscent of the Vienna Circle in his talk of 'factual statements', 'literal truth', 'literal meaning' and 'meaninglessness' (Hick, 1977b, pp. 177-8). But nowhere does he explain what he means by these terms.

Whether or not the Christian doctrine of the incarnation is soundly rooted in the early thought of the Church, and whether or not it is a sound doctrine, aspiring to the truth, are serious questions open to rational argument; but it is not good enough to reject the doctrine on the basis of the outmoded philosophical presuppositions of positivism.

Retrospect

Although Logical Positivism has had its day, it served a valuable purpose in its time. It was an attempt to achieve the impossible dream of Descartes — to provide sure and certain foundations for knowledge. The attempt raised a host of profound problems which have engaged the attention of subsequent philosophers of science. It is characteristic of the advancement of ideas and the growth of knowledge that a theory be forcibly propounded only to be equally forcibly rejected in order to clear the way for a better theory. Positivism cleared away much metaphysical verbiage with which philosophy was cluttered, and it made theologians more cautious about some of the more extravagant claims they once made. Its concern was to find a principle of demarcation between statements which really did say something and statements which did not. It has thrown up as many problems for science as it has for religion. And it has caused the philosophers of both science and religion to examine their respective concerns more closely and critically. Unfortunately the one is only superficially acquainted with the findings of the other. We have cited examples of theologians who, long after the demise of positivism in science, have deferred to its claims. Likewise there are scientists whose passing acquaintance with theology is, to say the least, somewhat medieval.

The presumptions of positivism lead to the profession of atheism. Our concern in this chapter has been to exhibit the nature of these presumptions and to indicate where they are fallacious. Further criticism of positivism will become evident as we proceed with alternative theories of science.

Popper relates that Frege, when faced with Russell's paradox, said 'Arithmetic is tottering.' He points out that Frege was wrong in his diagnosis. It was not arithmetic that was tottering, but the theory of arithmetic (Popper, 1979, p. 373). Likewise with science and religion. Many would hold that religion is tottering under the impact of science. But this is not so. It is only a particular *theory of science* that has such devastating consequences for religion. We do no service to either science or religion by uncritically accepting a particular theory of science and endeavouring to adapt the intellectual basis of religion to it.

Notes

1. There is no doubt that the authority and influence of science derive in the first place from the evident success of the applied sciences in their ability to ameliorate material conditions and improve man's lot. But our enquiry is concerned with the more fundamental intellectual underpinning, or rationale, of science rather than with the apparent advantages it confers in social and economic terms.

2. See Suppe's comment (1977, p. 6).

3. Putnam (1962) first introduced the term the *Received View*, and Suppe (1977) traces its development and subjects it to critical assessment.

4. The dispute regarding the import of Wittgenstein's thesis in the *Tractatus* is unresolved. It was Russell, in his introduction to the *Tractatus*, who suggested that Wittgenstein was concerned with the conditions of a logically perfect language. The main point here is that the Russellian model of analysis was used, even though, unlike Russell, Wittgenstein in the *Tractatus* does not refer at any point to sensations.

5. References in decimal notation are in accordance with Wittgenstein's notation in the *Tractatus*.

6. This distinction is discussed more fully by Parkinson (1968, Introduction) and Ayer (1976, Ch. 2).

7. Here is a significant, if previously unacknowledged, step towards the later Wittgensteinian identification of meaning with use, and it is a direct derivative of positivism. See Chapter 3.

8. The refusal to admit past errors is a characteristic of all authoritarian stances, in politics as well as in epistemology. It is also their fundamental weakness.

9. Bacon was not alone in his reaction to Aristotelian science. His contemporary Galileo, for instance (of whose work Bacon seems ignorant), was actually engaged in experimental work. Nevertheless, Galileo's great and original contribution to modern science lay in his use of mathematics and 'thought experiments'. The 'book of nature', for Galileo, was written in the language of mathematics.

10. In a later interview with Magee, Ayer still maintained that the book was 'true in spirit', and yet he makes the startling admission 'that nearly all of it was false' (Ayer, 1978, p. 131).

11. Ayer discusses this failure in *The Central Questions of Philosophy* (1976, Ch. 2).

12. See Carnap in 'Psychology in Physical Language' in which he states that 'psychology is a branch of physics' (Carnap, 1934).

13. Bultmann's demythologising involved a rejection of traditional religious talk and doctrinal formulation. Anti-natural (anti-scientific) miracles were allowed no place in his interpretation of the Gospels. He stressed the importance of personal response and the 'existential' character of faith. Hence the resurrection of Jesus was no longer an event in the physical world, but an event in the life of the faithful. The apostles' faith rose at Easter. 'The question of God and the question of myself are identical' (Bultmann, 1958, p. 53).

14. Love may well be the supreme virtue, but love, as a rationally unregulated emotion, can lead to disastrous consequences (see Ch. 2, 'The Implications for Religion').

15. Braithwaite was one of a number of philosophers of science of this period who produced a version of the Received View. Despite their individual, distinctive features, they are all vulnerable to the criticisms we have outlined.

16. Square brackets in original.

17. The extraordinary success and widespread appeal of *Honest to God* was in no small measure attributable to its sensitivity to the dilemma posed for those attempting to square religious faith with the Received View of science and its extension as a general epistemology.

18. Keith Ward subjects Cupitt's *Taking Leave of God* (1980) to rigorous and devastating criticism in *Holding Fast to God* (1982a).

2 THE RETREAT FROM AUTHORITY

Karl Popper[1]

It is generally agreed that the argument for Logical Positivism and the proclamation of the Verification Principle were introduced in Britain by A.J. Ayer in his book *Language, Truth and Logic* in 1936. Bryan Magee, in his BBC interview with Ayer in 1971, remarked that 'this was probably the most explosive philosophical work in English this century'. The book had a powerful influence and 'to this day there are many professional philosophers who, when asked to recommend an intro-ductory book by someone who thinks he might be interested in philo-sophy, recommend *Language, Truth and Logic*' (Magee, 1971, p. 48).

What is probably not generally recognised is that the arguments for Logical Positivism contained in *Language, Truth and Logic* were deva-statingly criticised, if not totally demolished, by Karl Popper in his book *Logik der Forschung* published two years earlier, in 1934. In other words *Language, Truth and Logic* was refuted before it was written.

Doubtless, the fact that this seminal book of Popper's on the philo-sophy of science was not introduced to an English-speaking public until 1959, under the title *The Logic of Scientific Discovery*, had much to do with this extraordinary state of affairs. Magee is probably right when he comments that 'the philosophy of a whole generation here in England might have been different if the book had appeared earlier' (Magee, 1971, p. 48).

The Logic of Scientific Discovery is probably the most important book on the philosophy of science to appear this century. Popper's radical and devastating criticism of positivism led him to reject the inductive logic on which it rested and to develop a fresh analysis of the logic and methodology of science. This in turn developed into a wider systematic philosophy, linking the growth of knowledge with biological and cultural evolution. What began as an attempt to characterise the logic and method of natural science developed into an evolutionary philosophy which embraces wide areas of human experience. This over-all sweep of Popper's philosophy is remarkable in the range and depth of the ideas involved, and it offers a refreshing and stimulating alterna-tive to those contemporary philosophers who, as heirs of the positivist tradition, have reduced philosophy to the dry bones of logic and the sterile analysis of language.

Although, as we shall indicate, there are serious difficulties in the Popperian analysis of the scientific enterprise, practically all the issues currently aired in the philosophy of science arise from a critical assessment of Popper's philosophy. Indeed, the new shift in the philosophy of science which was heralded by Thomas Kuhn, whose ideas will be discussed in the next chapter, arose as a reaction to, and a development of, Popperian philosophy of science.

Not only has Popper's work stimulated intense interest in the debate about the logic and method of natural science, but it has also affected the attitudes and working methods of many notable scientists in a way that no other philosopher of science has done.[2] What is more, Popper's work has shown that some understanding of the content and method of science is fundamental to philosophy in general. Ayer himself, who came to philosophy through a study of the classics, admitted later in life that he would have been a better philosopher had he had some grounding in science.

Popper has frequently complained that some of the criticisms of his work fall wide of the mark because his critics have often misread and therefore misunderstood his ideas. There is some justification in this complaint, as we shall indicate, and as Popper's thought is subjected to increasing critical attention, the commentaries will be more widely read than the original. Such is the lot of many great thinkers: the latter-day judgements are taken for the real thing, and some of the original insights are obscured.

In the preface to *The Logic of Scientific Discovery* Popper locates the central problem of philosophy as the problem of the growth of knowledge, and this in turn provides the key to the whole of his own wide-ranging thought. The best example of knowledge growing is provided by natural science. The positivists too located the problem of the growth of knowledge in natural science, but Popper parts company with them by providing a very different characterisation of science. His concern in this respect is to demarcate scientific knowledge from non-scientific knowledge and thus identify more clearly the characteristics of scientific knowledge and of rationality.

As we have observed, Hume's analysis of induction led him to designate science as an irrational activity, and although the positivists followed Hume's empiricist dogma that all our ideas are derived from our sense impressions, they were not prepared to accept his sceptical conclusions about the irrational nature of science. Much of their endeavour lay in their attempts to rehabilitate the logic of induction, for without it they had no way of characterising rationality.

Popper shares with the positivists the conviction that science is rational, but he turns the positivist programme on its head by substituting falsification for verification, theory for observation, fallibilism for certainty or high probability, and he locates the essential feature of rationality in criticism rather than in justification.

Although Popper began by examining the growth of knowledge in science, his inquiry develops into an evolutionary epistemology in which the growth of knowledge is seen as a continuous process with biological and cultural evolution. All organisms from the amoeba to man are constantly, day and night, engaged in problem solving. The critical-rational approach to knowledge is applied to society in general, and consistent with his rejection of justification in science is his attack on political authoritarianism which leads to totalitarian attempts to control society.

In his intellectual autobiography, *Unended Quest* (1976a), Popper writes about his early life in Vienna, at the same time indicating some of the rich cross currents of thought in the remarkable intellectual ferment of Vienna in the early years of this century. Doubtless this Austrian renaissance was in no small measure responsible for the wide-ranging span of Popper's interests and concerns, and the remarkable degree of originality and insight which permeates his thought. Popper's initial ideas were offered as a criticism and corrective to the ideas of the Vienna Circle. He claims to have 'killed logical positivism' (p. 87). The criticism proved fruitful and creative for in it were the seeds of most of his subsequent thought which he has refined and developed over the last half century.[3]

Popper was impressed by the logical formalism of the positivists, and he has retained an attachment to deductive logic as the touchstone of rationality throughout his life. But this logical rigour is presented in the context of a much wider sweep which is constantly informed by his realism, his sensitivity to the realities of scientific thought and practice, and above all by his concern for human freedom and the fact of suffering. His admiration for Bertrand Russell, the 'Passionate Sceptic', doubtless derives from these common concerns.

This is why Popper's ideas are stimulating and challenging. He does not limit philosophy to logical analysis, and he was never seduced by its successor which reduced philosophy to the analysis of language. The lifeblood of philosophy for Popper is problem solving. The problems are cosmological; concerned with understanding the world, including ourselves and our knowledge as part of that world. 'The phenomenon of human knowledge is no doubt the greatest miracle in our universe' (1979, Preface). This is his canvas. The ideas are bold; they are certainly open to objection and criticism, but through the development of these ideas Popper has helped to enlarge our vision, in contrast to many of his philosophical peers with their myopic obsession with logical and linguistic minutiae.

Popper did not set out to produce an all-embracing philosophy after the style of the great metaphysical system builders of the past. Rather, he proceeded in the tradition pioneered by Russell of piecemeal analysis of separate problems. But as John Watkins points out in his essay 'The Unity of Popper's Thought' (1974, Ch. 11), something like a systematic philosophy emerges as a result of connections and unifying ideas that link his thought in different fields.

Because our task is also a wide-ranging one, embracing philosophy of science and philosophy of religion, we must treat Popper as a systematic philosopher and consider his overall views and their implications rather than concentrate only on the logic and methodology of science to which he first addressed himself. Although Popper has elevated criticism above all other intellectual and practical virtues, the most rewarding and fruitful way to understand his contribution to philosophy, especially with regard to its implications for religion, is to attempt to assess his work positively and to develop his insights. Such an assessment will provide the best context for criticism.

As we have indicated, Popper's enduring concern is with the *growth* of knowledge. Growth implies change, and this brings us to the source of the great fertility of Popper's ideas: his distinction between the two kinds of change which might occur among things and among human constructions. Popper has characterised this distinction as between Darwinian selection and Lamarckian induction. His insistence on the Darwinian selection-elimination model is at the heart of all his philosophy — of science, of history, of society and of metaphysics, indeed of all rationality.

Our procedure then, will be to begin by citing the Darwinian-Lamarckian distinction in biology and tracing this idea as Popper develops it into an evolutionary epistemology, a theory of the growth and expansion of knowledge.

Such a procedure involves a departure from currently accepted philosophical methods, for it involves offering what D.T. Campbell has called a 'descriptive epistemology' — that is, descriptive of man as knower (Campbell, 1974, p. 413). Such a description involves a high degree of empirical content and biological information. It will, of course, be objected that this is not the domain of philosophy but of the respective natural sciences; but all philosophical theories are dependent on a concealed descriptive epistemology. The philosopher must have something to work on. The positivist descriptive epistemology was based on induction, while that provided by Kuhn, for instance, is a historico-sociological account. Two of the great mistakes in philosophy, which positivism inherited, resulted from uncritically accepting faulty descriptive epistemologies. One was Locke's theory of knowledge, based on the *tabula rasa* theory of the mind; the other error, which its author lived to refute, was Wittgenstein's attempt to characterise language, in the *Tractatus*, without taking into account any notion of its origin and evolution and diversity of functions. In his preface to that powerful and influential book Russell commended its impeccable logic without questioning the empirical assumptions on which it was based. But as Campbell points out, an epistemology must not only be analytically consistent but also compatible with the description of man and the world provided by contemporary science. It is to such a description that we now turn.

The Evolution of Persons

The Evolutionary Continuum

If we are to examine the growth of knowledge in its evolutionary context, we must begin by identifying three broad phases of evolution: inorganic, organic and human. *Inorganic* evolution is the domain of physics and chemistry in terms of which the evolution of the elements that constitute the universe is accounted. The marriage between cosmology and particle physics in contemporary physical theory is proving most fascinating and fruitful. The time scale involved here takes us back to the Big Bang, some 15 billion years ago. *Organic* evolution is concerned with the evolution of life from primitive unicellular structures to the staggering complexities of the human brain. In this realm we are dealing with the last 3 to 4 billion years. *Human* or *cultural* evolution is of much more recent origin; but this third phase, of human life, marks the emergence of conscious, sapient, purposive life and the beginning of man's cultural history in which language, knowledge and understanding develop conjointly.[4] Any philosophy of science which deals with physics, biology and the human sciences must take cognisance of these phases of evolution. A descriptive epistemology is particularly concerned with biological and human evolution.

Emergence and Reduction

When we consider the evolution of life on this planet we notice an important feature — the emergence of new forms: the emergence of life from inanimate matter, the emergence of consciousness from life, and the emergence of language, knowledge and understanding from consciousness. There is an important point here with regard to 'reductionism'. The thesis of physicalism, and in some of its aspects, behaviourism, dogmatically asserts that because man, with conscious, purposive experience has evolved from inanimate matter, then he can be reduced, without remainder, to physical descriptions. In Carnap's words, 'psychology is a branch of physics' (1934). (Or as Greta Garbo once put it, 'Love is just a chemical reaction'!) It is central to Popper's whole philosophy that evolution is an indeterminate, unpredictable affair, in which new, emergent forms are thrown up which can only be understood on their own terms and not simply in terms of the previous forms from which they have emerged. An emergent form, although having evolved from a previous, less complex, form, cannot be reduced to the previous form from which it has emerged. Although Popper is scornful about philosophical reductions, which are totally unwarranted, he does not underrate the value of attempted scientific reductions: of chemistry to physics, of biology to chemistry, of psychology to biology, of sociology to psychology. Each of these forms a valuable research programme which helps to increase our understanding of the world. Nevertheless, he argues in 'A Realist View of Logic, Physics and History' (1979, Ch. 8), and at greater length in *The Open Universe* (1982b), not only that no completely successful reduction has ever been achieved, but that it is doubtful if it ever could be achieved. His main argument here is that at each higher level there are developments which could not have been foreseen at lower levels. This is why reductionism and determinism are linked theses. His particular concern is with the emergence of consciousness, which, he argues, could not have been potentially present in fundamental physical particles (quarks?). Had they been, then sufficient knowledge of these potentialities would allow the prediction of life and consciousness beforehand.

Darwinism and Natural Selection

The breakthrough in evolutionary theory attributed to Darwin is contained in the idea of natural selection. We need to examine this notion in some detail because of its implications, not only for biological evolution, but for its relevance to evolutionary epistemology, a theory of the growth and expansion of knowledge.

Darwinian natural selection can best be understood in contrast to the Lamarckian account of evolution which was popular in the early nineteenth century and which the layman even today finds intuitively attractive. According to Lamarck the environment induces a change or an improvement in an organism. New organs arise from new needs, and they develop in proportion to the extent to which they are used, the new acquisitions being handed down from one generation to the next. Conversely, disuse of existing organs leads to their gradual disappearance. Hence, according to Lamarck, it is the environment that acts directly, or indirectly, on the individual organism, and new characteristics thus acquired by the individual are handed down to its descendants. The organism thus possesses an inherent faculty for self-improvement: the giraffe, for instance, gets a long neck because generations of giraffes have stretched to reach the higher foliage.

Darwin, in *The Origin of Species* (1902, Ch. 4), postulated that the main cause of modification in a species is due to variations that spontaneously arise in the passage from one generation to the next. The cause of these variations was unknown to Darwin, but if they happen to be useful they will survive by natural selection. Natural selection operates because organisms tend to multiply by geometrical progression while the means of subsistence cannot do so. Consequently there is a weeding out in each generation. The least fit perish, and any random variation which promotes the chance of survival means that the possessor of the favourable variation will hand it down to the next generation. The accumulation of favourable variations over long periods will result in the emergence of new species and the extinction of older, less well-adapted species. Natural selection operates over the whole organic realm from the amoeba to man.

The neo-Darwinian synthesis has combined Darwin's fundamental insight regarding natural selection with an increasing understanding of genetics, from Mendel's epoch-making experiments with peas in 1866 to the brilliant unravelling of the DNA molecule by Crick and Watson in 1953. Jaques Monod summed up the process in the words of Democritus, 'Everything in the universe is the fruit of chance and necessity' (1972, p. 114). Blind or random mutations of genes are faithfully replicated and the variations produced are controlled by natural selection. No external influence can affect the genome.

Popper does not accept Darwinism as a testable scientific theory because of its near tautological nature — the best fitted to survive will survive — and its lack of testability. However, it does provide what Popper calls 'a metaphysical research programme' and there is no doubt that it has stimulated biological research with remarkable effect during the last hundred years. Its further importance for Popper is that it has provided him with a general framework for the development of his ideas.

Problems, of course, remain for the Darwinian account of evolution, particularly with regard to the *rate* of evolutionary change, and the suggestion of direction or purpose in the evolution of complex organs such as the eye.[5] Popper himself has attempted to resolve the question of direction or goal-orientation in evolution by suggesting a genetic dualism rather than genetic monism. That is, he distinguishes between different classes of genes: those controlling an organism's preferences, aims and skills; and those controlling its anatomy or structure. Although all genetic changes are still random on this view, the preference and skill levels control the selection of the changes at the anatomical level, so we can see why such changes as appear tend to serve the aims of the organism. Thus the organism is viewed as a hierarchical system in which there is interplay and feedback between the various levels of control.

Indeterminacy and Plastic Control

Such a hierarchical genetic system implies what Popper describes as a system of 'plastic controls'. He maintains (contrary to Hume) that there is a middle way between sheer chance on the one hand and complete causal determination on the other: between Monod's chance and necessity. Popper develops this theme of indeterminacy and plastic control in 'Of Clouds and Clocks' (1979, Ch. 6). The whole theme of indeterminism is treated at length in *The Open Universe* (1982b).

Having argued the case for physical indeterminism he maintains that 'indeterminism is not enough'. Sheer randomness is no more satisfactory than determinism because both rule out purpose and deliberation and conscious action. If we are to understand rational human behaviour we need something between chance and necessity; some level of control between the random movements of the molecules in a gas cloud and the rigidly determined structure of a clock. He demonstrates the notion of plastic, or soft, control in physical systems with the example of a soap bubble which consists of two intermediate systems which control each other. Without the air the soapy film would collapse, and without the soapy film the air would be uncontrolled. 'Thus control is mutual: it is plastic, and of feedback character' (1979, p. 249). There is a degree of give and take between the outer film and the inner molecules. This notion of plastic control standing between 'iron' control and no control is applicable not only in the realm of physics but also in that of biology. A hierarchical system, such as a complex organism, viewed as a unitary whole, is far more determinate than the behaviour of its component parts. But both complete determinism and sheer randomness are ruled out in view of the give and take nature of plastic control.

The Emergence of Consciousness

These notions of indeterminism, genetic dualism and plastic control allow us to conceive of an organism as a quasi-hierarchical system of plastic controls. Somewhere along the evolutionary trail consciousness emerges, and in its turn acts with increasing effect within the control mechanism of the organism. Human self-consciousness is fundamentally irreducible to any lower level of existence. Indeed, Popper sees it as the basic immaterial characteristic of the Cartesian self; he does not shrink from the Cartesian notion of a mind-body dualism which his depiction of conscious control implies. The 'mind-body liaison', as Eccles the neurophysiologist puts it, is for Popper one of the miracles of evolution which will probably for ever defy complete human understanding.

In *The Self and its Brain* (Popper and Eccles, 1977), Popper deploys ingenious arguments against epiphenomenalism and the identity theory: his point being that if consciousness has emerged under the rigours of natural selection it must confer some advantage on the organism. This would be denied if consciousness, or mind, is simply identified with physical processes, or if it were a mere accompaniment (or epiphenomenon) of existing physiological processes. Hence, the existence of conscious control is of crucial importance in Popper's descriptive epistemology.

Problem Solving

The view of evolution so far developed here implies a growing hierarchical system of plastic controls, of which consciousness is not only one of the latest to develop but also one of the most effective. This system of plastic controls is incorporated genetically in the organism, and in the case of man, as we shall see, plastic controls evolve more and more exosomatically. Mutations are seen as trial and error gambits and natural selection as a way of controlling them by error elimination. This brings us to Popper's central thesis, which informs the whole of his epistemology, and which is at the heart of his philosophy of science:

All *organisms* are constantly, day and night, *engaged in problem solving*; and so are all those evolutionary *sequences of organisms* – the *phyla* which begin with the most primitive forms and of which the now living organisms are the latest members. (1979, p. 242)

These problems are objective in the sense that they can be constructed by hindsight, and their solution proceeds by the method of trial and error – tentative solutions are controlled by error elimination. This gives rise to a new problem situation which in turn invites tentative solutions. Popper schematically describes this sequence of events as follows:

$$P_1 \rightarrow TS \rightarrow EE \rightarrow P_2$$

where P is the problem, TS the trial solution, and EE error elimination. Many trial solutions are put forward and the resulting problem situation P_2 is different from the one that gave rise to it. The process is not cyclic, but one in which feedback, from P_2 to P_1, takes place.

Thus the selective elimination model of biological evolution which throws up problems for solution is a knowledge process. A physical system, like a soap bubble, has no problems, but with the amoeba the case is different. It has food problems, and is constantly engaged in an attempt to ingest its immediate environment. It is this feature of problem solving that distinguishes the biological from the merely physical; problem solving, and hence natural selection, is the essential characteristic of life. Viewed in this way, an organism's 'blind' movements can be regarded as a series of trial solutions to the problem that confronts it. At the lower level, a mutation is a trial solution to a problem presented by the environment. At the level of more complex organisms viewed as a unitary whole, the control element can be seen as the selective capacity for rejecting unsatisfactory trial solutions and pursuing more rewarding ones. The solution to one problem creates a fresh problem situation, and so on. The problem solving process only ends when the organism reverts to the problem-free situation of dead matter, that is, when it dies.

The evolutionary hypothesis implies a fundamental indeterminism. The rigid (Laplacean) determinist would hold that, given enough information about initial conditions, the whole course of evolution to the present and beyond could be predicted. This analytic approach cannot be accommodated in the Darwinian selective-elimination model we have outlined. To the theme of indeterminism we shall return.

It might be objected that we can only speak of problem solving metaphorically, because evolution is clearly not a conscious process. But who has decreed that we should not speak of problems in an objective, non-psychological sense? As Popper points out, when we speak of a problem we do so almost always with hindsight. We may work on problems, the precise nature of which are unclear, and we very often mistake our problems. Even in the field of science in which scientists consciously try to be aware of their problems, there are many problems tackled and solved unconsciously. Thus we may say that the amoeba solves problems as does the scientist, and we need not assume that either is in any sense aware of its problems.

Popper sees Darwinism as an application of what he calls 'the logic of the situation' (1976a, p. 168). That is, the manner in which a problem is tackled will depend on the constraints imposed upon it by virtue of the conditions prevailing at the time. The logic of the situation is Darwinian rather than Lamarckian: that is to say, the response to the problem is creatively thrown up as a hypothesis which is tested by environmental pressures which lead to its complete rejection, modification, or acceptance.

Thus the evolutionary process may be viewed as one in which an organism 'learns' to organise its experience of its environment. At the level of the lower organisms this is an unconscious process, the amoeba searching for food. At the highest level we have the scientist consciously searching for an understanding of the world with his theories. At the biological level successful organisation of experience is necessary for the survival of a species, and it determines the ecological niche into which a particular organism will fit. The ant and the oyster, for instance, have so successfully adapted to their highly specialised environment that they have remained virtually the same for millions of years. In contrast, one of the distinguishing features of man's evolution is that he has become to a large extent emancipated from dependence on his immediate environment, and his evolution has been comparatively rapid in his recent evolutionary history. He spread initially from the equatorial area to the polar regions, and more recently gained a foothold on the inhospitable moon.

It is at this point in man's biological evolutionary history that his evolution is not only endosomatic, but becomes increasingly exosomatic. Both endosomatic and exosomatic 'organs' develop conjointly, and as the exosomatic evolution proceeds, so it increasingly plays an important part as a 'plastic' control. The increasing role of man's exosomatic evolution brings us to the third phase of evolution, the cultural, or human. Language is clearly the key to this remarkable and rapid development, in which man becomes increasingly emancipated from his dependence on his immediate environment, and it provides the link between man's biological and cultural evolution. It is the crucial step that divides the amoeba from man.

The Emergence of Language – the Genetic Factors

The emergence of language marks the third phase of man's evolution, the human and the cultural. An adequate descriptive epistemology must above all include an account of the functions of human language in an evolutionary context. Because language forms the bridge between biogical and cultural evolution, its development is dependent on both endosomatic and exosomatic factors. Language was, of course, at first *spoken*, and this remains its prime function, but the invention of writing marked a step of immense significance in man's cultural evolution. What was previously encoded in artefacts and objects and stored in the human memory and passed on by oral tradition now became encoded in written language. Not only could information, thoughts and ideas be disseminated in one age, but they could be recorded for recovery in later ages.

The remarkable development of the human cortex over the span of a mere 2 million years suggests that the emergence of language greatly increased the survival value of intelligence, thus conferring considerable selection pressure in favour of the brain. Monod suggests:

> The day when Australanthropus or one of his kin ventured beyond communication of actual experience and expressed a subjective experience, a personal 'simulation', saw the birth of a new world, the world of ideas: and a new evolution, that of culture became possible. From there on man's physical evolution was to continue for a long time, closely related to the evolution of language, and deeply subjected to its influence, which so changed the conditions of selection. (1972, p. 150)

This suggests that man's language ability is genetically encoded, rather than inductively developed by imitation, and it has provided the spur for the evolution of the brain and therefore of human intelligence. Monod, once again, puts it rather nicely when he says:

> the day when Zinjanthropus or one of his friends first used an articulate symbol to represent a category, he immensely increased the pro-

bability that a brain might one day emerge capable of conceiving the Darwinian theory of evolution. (1972, p. 130)

This conclusion is supported by Chomsky's linguistic researches, and his conclusion that human beings are genetically endowed with a highly specific language faculty: that a child has *inborn knowledge* of the universal principles governing the structures of human language. Popper is fond of supporting this thesis with the example of Helen Keller, who, although blind and deaf from birth, appeared to have an innate, inborn disposition to organise her experience with the use of language.

Popper, who, as we have indicated, is both a dualist and an interactionist, shows further affinity with Descartes when he suggests that the physiological basis of the human mind might be found in the speech centre of the brain, and that this might locate the seat of consciousness (1976a, p. 189, p. 237 note).

The Emergence of Language – the Cultural Factors: an Evolutionary Sequence

This brief excursion into the biological origins of the evolution of language needs to be supplemented by an account of the cultural evolution of language; that is, how man has developed the use of language, and how it in turn has affected his social and intellectual development. It is here that Popper offers a theory of the *functions* of language which helps us understand the significance of human language. Popper is highly critical of the linguistic philosophers who, as heirs of the positivistic tradition, direct our attention to the meaning of words, which they identify with their use, rather than to problems vicariously represented in language. If the descriptive epistemology, which regards language as part of evolution as a problem solving and knowledge process, is correct, then we need to look beyond the analysis of linguistic usage to the functions of language within the evolutionary context. Popper likens language to spectacles. Although you clean your spectacles they only function when you put them on and look through them at the world.

It is the same with language. That is to say, one shouldn't waste one's life in spectacle cleaning or in talking about language, or in trying to get a clear view of our language, or of our 'conceptual scheme'. The fundamental thing about human languages is that they can and should be used to describe something; and this something is, somehow, the world. To be constantly and almost exclusively interested in the medium – in spectacle cleaning – is a result of a philosophical mistake. This philosophical mistake can be quite easily traced in Wittgenstein. Wittgenstein was originally impressed by the fact that the medium may impose limitations on us, or that it may actually deceive us, and he was also impressed, upon further analysis, by the fact that we can't really *do* anything about the medium. By a kind of recoil from his earlier views he then said that all we can really do is to list the various usages (of a word, for example), to try to see the differences, and to understand them somehow. The real issue as I see it is different; do we have a philosophy of language which explains to us the functions of language, and which helps us to understand the significance of human language (which is more than a game)? (1971, pp. 138-9)

Popper develops a theory of Karl Buhler which attempts to trace the evolution of language from animal language to human languages. He identifies two lower functions which animal and human language share; these are the *symptomatic* or *expressive* function, and the *releasing* or *signalling* function. An organism expresses its state with linguistic signs, and this in turn becomes a signal, a means of communication, when the receiving organism responds to it. These two lower functions are a characteristic of all linguistic phenomena, and they are *always* present in human language. But human language is characterised by other functions, the most important of which are the *descriptive* and the *argumentative* functions. These two higher functions of human language are coupled with the evolution of reasoning and rationality; with the development of knowledge and the growth of science.

As an example of the descriptive function, I might now describe to you how two days ago a magnolia tree was flowering in my garden, and what happened when snow began to fall. I might thereby express my feelings, and also release or trigger some feeling in you: you may perhaps react by thinking of *your* magnolia trees. So the two lower functions would be present. But *in addition* to all this, I should have described to you some facts; I should have made some *descriptive statements*; and these statements of mine would be factually *true*, or factually *false*. (1979, p. 236)

It is with the development of the descriptive function that language becomes characteristically human and it marks the emergence of man from the animal kingdom. The formulation of descriptions of the world, of actual or possible states of affairs, leads to understanding and the concepts of truth and falsity. As human language developed in this way so man became self-conscious: human language and human consciousness are interdependent and develop conjointly. The first use of descriptive language was for true reports of states of affairs, then came the invention of lying and with it the notion of falsity and the development of story telling. Every genuine report contains an element of decision: the decision to tell the truth. Experience with lie detectors indicates that the propensity to speak the truth is biologically earlier than that of lying (1974, p. 1113). The idea that truth has a biological origin is one that has been overlooked by philosophers. If man has an inborn need for finding regularities, and inborn mechanisms of discovery which make him search for regularities, then he also has an inborn need to communicate true reports and descriptions, otherwise his experience would be chaotic and disorganised. In this sense truth has enormous social significance. R.D. Laing and A. Esterson (1970) have given a plausible account of how schizophrenia in one member of a family can result from a calculated campaign of deception in which false accounts and interpretations of events are consistently presented by other members of the family.

The fourth, and highest, function of human language is the *argumentative* function.

> The argumentative function of language is not only the highest of the four functions . . . but it was also the latest of them to evolve. Its evolution has been closely connected with that of an argumentative, critical, and rational attitude; and since this attitude has led to the evolution of science, we may say that the argumentative function of language has created what is perhaps the most powerful tool for biological adaptation which has ever emerged in the course of organic evolution.
>
> Like the other functions, the art of critical argument has developed by the method of trial and error-elimination, and it has had the most decisive influence on the human ability to think rationally . . . like the descriptive use of language, the argumentative use has led to the evolution of ideal standards of control, or of *regulative ideas* (using a Kantian term): the main regulative idea of the descriptive use of language is *truth* (as distinct from *falsity*); and that of the argumentative use of language, in critical discussion, is *validity* as distinct from *invalidity*). (1979, p. 237)

Critical argument has a direct bearing on the evolutionary theme of problem solving: problems are objectively formulated in language, particularly in propositional form, and very often arise out of descriptive propositions. Language can function descriptively without the argumentative function, but when reasons are given for or against the truth of descriptive statements, the argumentative function comes into play.

Linguistic Subjectivism

There has been a tendency, evident in traditional epistemology, to overlook the higher forms of language because they are always accompanied by the two lower functions. Consequently it is always possible to interpret every linguistic phenomenon in terms of the lower function as an *expression* or *communication*. But theories of human language that focus on expression and communication lead to a subjectivist epistemology and a behaviourist psychology. If we do not take into account the descriptive function of language, then our notion of *correct* description or of objective truth, becomes redundant: truth becomes relative rather than an objective, regulative principle. Likewise, neglect of the argumentative function of language is equivalent to a rejection of criticism, and this in turn leads to intellectual stagnation rather than evolution and growth of knowledge. Popper is scathing in his condemnation of this attitude in his essay 'Towards a Rational Theory of Tradition', when he refers to the new enemies of reason and their rejection of the rational, critical tradition in certain types of modern poetry, prose and philosophy – 'philosophy which does not argue because it has no arguable problems'. They uphold implicitly a theory of language that sees no more than the first or perhaps the second of its functions, while in their practice they support the flight from reason and from the great tradition of intellectual responsibility' (1972, p. 135).

Learning Language by Trial and Error

Our evolutionary thesis implies that language has evolved both endosomatically and exosomatically by trial and error. Hence there are no determinate meanings and it is important to recognise the approximate and pragmatic character of language at all stages in its evolution. Attempts at definition lead to an infinite regress and certainly cannot be founded on language. D.T. Campbell, in his essay 'Evolutionary Epistemology', describes how a child learns a language by trial and error usage, rather than by an inductive process of passively observing adult usage. Word meanings are not directly transferred to the child; rather, the relationship between things and actions and words is a contingent one; nothing is entailed, and it is always approximate. Sets of ostensive instances are always incomplete and allow manifold interpretations. This trial and error process of learning language requires more than the communication of teacher and child; it requires a third party of objects referred to. 'Language cannot be taught by telephone but requires visually or tactually present ostensive referents stimulating and editing trial meanings' (Campbell, 1974, p. 433).

Language, Plastic Control and Feedback

Language is high in the hierarchy of 'plastic controls'. Linguistically formulated conjectures, theories and aims can affect our actions and movements, if we freely submit to them. It is in this sense that Popper offers a solution of what he calls 'Compton's Problem' by demonstrating how abstract meanings can control our behaviour. But the control is far from one-sided for we can control our theories and reject them after critical discussion (see 1979, pp. 240-1).

We have noted that biological evolution proceeds largely by the modification or emergence of new organs, and that human evolution proceeds largely by developing exosomatic 'organs' or artefacts. Significant examples of this exosomatic evolution are tools, weapons, farming implements, the wheel, fire and more recently computers[6] and space vehicles. One of the most important examples of exosomatic evolution is that of writing, the advent of which marked a critical turning point in man's cultural evolution. Man, having become articulate, becomes able to record his thoughts and imaginings and ideas. With regard to this capacity for invention, Popper observes:

> man, instead of growing better eyes and ears, grows spectacles, microscopes, telescopes, telephones and hearing aids. And instead of growing swifter and swifter legs, he grows swifter and swifter motor cars . . . instead of growing better memories and brains, we grow paper, pens, pencils, typewriters, dictaphones, the printing press, and libraries. (1979, p. 238)

Not only are these inventions a consequence of our use of language, but they have a feedback effect on language in the sense that they add new dimensions, especially to the descriptive and argumentative functions of language.

Conclusion

Hence the higher functions of language have made reasoning and understanding possible; they are at the root of the power of human imagination and inventiveness. Criticism, which arises with the argumentative function of language, becomes the main instrument for further growth. The schema

$$P_1 \rightarrow TS \rightarrow EE \rightarrow P_2,$$

originally valid for the animal world as well as for primitive man, becomes the schema of the growth of knowledge through error elimination by way of systematic rational criticism. Man has used language to create a body of objective knowledge. Knowledge is an evolutionary, exosomatic artefact which enables man to profit from the trials and errors of his ancestors. From this point onwards, the evolution of man is the evolution of his knowledge, and it is to this human knowledge that we must now turn.

The Evolution of Knowledge

Evolutionary Epistemology[7]

We have noted that the characteristic feature of biological evolution according to neo-Darwinism is contained in the idea of natural selection. This is in direct opposition to Lamarckian evolutionary theory, according to which the organism is instructed from without by the environment. Lamarckism involves a biological inductive process of learning by experience. Constant repetition of sense experience imprints behavioural characteristics on the organism and the characteristics thus acquired are passed on to its descendants. The accumulation of similar, externally derived impressions, implanted in successive generations, leads to the evolution or development of new organs which more effectively adapt the animal to its environment. Neo-Darwinism, in contrast, involves no such inductive process of learning by experience. Rather, the blind, or chance, mutations within the organism are tested and selected by the environment, and in accordance with the organism's hierarchical control system. Favourable mutations enhance the animal's skill in coping with its environment, thus increasing its survival potential.

D.T. Campbell, in his essay 'Evolutionary Epistemology', refers to this selective elimination process as 'an opportunistic exploitation of coincidence' in which there is no logical necessity, no absolute ground for certainty, but instead a most back-handed indirectness . . . At no stage has there been any transfusion of knowledge from the outside, nor of mechanisms of knowing, nor of fundamental certainties . . .

Considered as improvements or solutions, none of these variations has any *a priori* validity. None has the status of revealed truth nor of analytic deduction. Whatever degree of validation emerges comes from the differential surviving of a winnowing, weeding-out process. (1974, pp. 414-15)

This natural selection paradigm is imported by Popper into his evolutionary account of the growth of human knowledge. It makes its first appearance in his seminal book, *The Logic of Scientific Discovery*.

According to my proposal, what characterises the empirical method is its manner of exposing to falsification, in every conceivable way, the system to be tested. Its aim is not to save the lives of untenable systems but, on the contrary, to select the one which is by comparison the fittest, by exposing them all to the fiercest struggle for survival. (1968a, p. 42)

How and why do we accept one theory in preference to others? The preference is certainly not due to anything like an experiential justification of the statements composing the theory; it is not due to a logical reduction of the theory to experience. We choose the theory which best holds its own in competition with other theories; the one which, by natural selection, proves itself the fittest to survive. This will be the one which not only has hitherto stood up to the severest tests, but the one which is also testable in the most rigorous way. A theory is a tool which we test by applying it, and which we judge as to its fitness by the results of its application. (1968a, p. 108)

Popper's concern in *The Logic of Scientific Discovery* is with the *evolution* of knowledge, with its growth and expansion, and in particular with the evolution of *scientific knowledge* which, he maintains, is common-sense knowledge writ large. 'One may discern something like a general direction in the evolution of physics — a direction from theories of a lower level of universality to theories of a higher level' (1968a, p. 276). This evolutionary theme is manifest in the title and content of a remarkable book by Einstein and Infeld, *The Evolution of Physics* (1961), which portrays the evolution of ideas and theories in physics with a beautiful and comprehensive clarity. It was in fact the theories of Einstein, in contrast to the theories of Freud, Adler and Marx, that led the young Popper to develop his criterion of demarcation by empirical falsification as the characteristic of science (1972, pp. 33-7).

The Popperian Account of the Growth of Knowledge

Conjecture and Refutation. We have examined the traditional, positivist account of the way (scientific) knowledge advances, by observation and induction. The rationale of science, and therefore of the growth of knowledge, provided by this inductive account is logically the same as the Lamarckian account of evolution. Common to both is the notion that it is the environment that instructs us. Popper totally rejects induction and replaces it by falsification as the critical method of science, as Darwin had previously rejected the inductivism of Lamarck and replaced it by natural selection.

Popper's thesis rests on the logical asymmetry that exists between verification and falsification. Although theories cannot be verified, because of their unrestricted generality, or universal nature, they can be falsified by a single negative instance. He reformulates Hume's problem of induction, not in terms of our subjective beliefs, but objectively, as a problem of the logical relationship between singular statements and universal theories. In this form the problem of induction can be bypassed: universal theories are not deducible from singular statements, but they may be refuted by singular statements since they may clash with observable facts.

This treatment of the *logical* problem of induction gives rise to a new theory of the *method* of science; to an analysis of the *critical method*, the method of trial and error. Theories are not formed by inductive generalisation, nor can they be verified by heaping up positive confirmation. They start as imaginative conjectures, and the aim of observation is to try to falsify them by discovering the falsity of the observable consequences that can be deduced from them.

The preoccupation of the positivists with definition and linguistic precision, labelled by Popper *Essentialism*, is a self-defeating and trivial effort. It is self-defeating because definition is an unending process, and it is trivial because it substitutes verbal discussion for testable problems. The scientific attitude is marked by a determination to test severely and falsify theories, and not to attempt to produce verbal mirror images of reality. We accept theories tentatively if we fail to falsify them, but their subsequent elimination always remains a possibility. Einstein's theory of gravitation, which replaced Newton's theory, remains, to date, unfalsified; but it could in turn well be replaced by a better theory in the future. And that, indeed, was Einstein's attitude; he never conceived of his theories as anything but tentative steps in the right direction. The basic test statements which are deduced from the theory as potential falsifiers are not incorrigible statements about sense experience, but are statements about publicly observable things which are themselves testable, in the light of dependent theory. Thus these test statements have no ultimate authority; they are only provisionally and conventionally accepted until a decision is made to test these in turn. The admission of decision making and agreement in science introduces the idea of the *social* character of science which has become increasingly important in subsequent attempts to characterise the scientific enterprise.

Knowledge, then, is not accumulated by inductive inference from a passive, Lockeian type of indubitable sensation. Rather, it is conjectural and theoretical; a matter of putting imaginative questions to the world and actively seeking a negative answer. It is a matter of trial and error, of conjecture and refutation. Our conjectures about the world are not the direct results of our observations because our observations are theory-soaked: every observation is made in the light of a theory. Knowledge precedes observation. Rather, our conjectures about the world are free creations for which we have a natural, in-built biological propensity. Science begins when we consciously adopt a critical attitude towards our theories which we try to eliminate by subjecting them to severe tests. Herein lies the crucial difference between Einstein and the amoeba: Einstein consciously seeks for critical error elimination from his objective, linguistically formulated theories; but the amoeba cannot be critical because its hypotheses are part of it. The amoeba perishes with its false theories, but Einstein survives his false theories which die in his stead.

Thus the general picture of science is: we choose some interesting problem. We propose a bold theory as a tentative solution. We try our very best to criticise the theory; and this means that we try to refute it. If we succeed in our refutation, them we try to produce a new theory, which we shall again criticise; and so on. In this way, even if we do not succeed in producing a satisfactory theory, we shall have learned a great deal; we shall have learned something about the problem. We shall know where its difficulties lie. The whole procedure can be summed up in the words: bold conjectures, controlled by severe criticism which includes severe tests. And criticism, and tests, are attempted refutations. (1971, p. 73)

Although the aim of science is truth, we can never lay claim to it. Our knowledge remains forever provisional and conjectural because the best tested theory may turn out to be false. But verisimilitude, or approximation to the truth, increases as more and more false hypotheses are eliminated. The quest for verification and proof, or high probability on the part of the positivists, is replaced by ongoing criticism and fallibilism. The growth of our knowledge, then, does not proceed by accumulating secure evidence, but from problems and our attempts to solve them by subjecting them to severe criticism. This critical method is at the heart of rationalism.

Demarcation. The principle of empirical falsification, which is central to Popper's evolutionary epistemology, directly relates to the problems of demarcating empirical science from non-science and metaphysics, and of establishing the conditions of acceptability of scientific theories. A theory is empirical, or scientific, to the extent that it is open to empirical falsification: that is, if there is 'a possibility of a clash with test statements' (1974, p. 987).

It is important to recognise that Popper's criterion of demarcation is *not* a criterion of meaning, as was the Logical Positivist verification criterion. Unfalsifiable philosophical, or metaphysical, theories are not meaningless. The positivists, who were so obsessed with the metaphysical concept of 'meaning', persisted for many years in misinterpreting Popper's criterion of demarcation in terms of meaning, as Ayer, for instance, now ruefully admits (Ayer, 1959, pp. 13-14). Science itself often arises out of myths, metaphysical conjectures and dogmas handed down by tradition. What is more, it is not always possible to eliminate all metaphysical elements from scientific theories. All theories say more than we can test, and often they contain undefined or blurred concepts such as 'force', 'field' and 'particle'.

Nevertheless, non-testable, irrefutable metaphysical theories, not least the philosophical theories contained in this book, may be rationally arguable and critically assessed. In his discussion of 'The Problem of the Irrefutability of Philosophical Theories' (1972, Ch. 8), Popper maintains that central to the discussion is the problem situation or context in which the problem is set; or, we would add, the relevant descriptive epistemology. Popper's discussion of indeterminism, realism and objectivism in the *Postscript*[8] provides examples of rationally arguable metaphysical theses.

Popper has never denied the difficulties arising from his demarcation criterion, but he sees it as central to his whole philosophy. The reason for this is that he views criticism as the heart of rationality and therefore 'the demarcation between science and metaphysics is a special case of the wider problem of demarcating criticisable from non-criticisable theories' (1968b, p. 95). He admits that his theory of scientific method hinges on this issue of demarcation, it is 'the centre of the dispute', and it is here that his ideas are most severely challenged (1974, pp. 976ff).

Subjective and Objective Knowledge. Popper's account of the growth of human knowledge breaks with traditional epistemology by drawing a distinction between subjective and objective knowledge. It is his contention that

> traditional epistemology has studied knowledge or thought in a sub-jective sense — in the sense of the ordinary usage of the words 'I know' or 'I am thinking'. This, I assert, has led students of episte-mology into irrelevancies: while intending to study scientific know-ledge, they studied in fact something which is of no relevance to scientific knowledge. For *scientific knowledge* simply is not know-ledge in the sense of the ordinary usage of the words 'I know.' While knowledge in the sense of 'I know' belongs to . . . the world of *subjects*, scientific knowledge belongs to . . . the world of objective theories, objective problems, objective arguments . . . The traditional epistemology of Locke, Berkeley, Hume and even Russell, is irrele-vant, in a pretty strict sense of the word. It is a corollary of this thesis that a large part of contemporary epistemology is irrelevant also. (1979, p. 108)

'The traditional theories of knowledge (from Plato's *Theaetetus* to Witt-genstein's *On Certainty*) fail to make a clear distinction between object-ive and subjective knowledge' (1974, p. 1028). They result from a 'sub-jectivist blunder' which has dominated Western philosophy (1979, Preface).

According to the common-sense theory of knowledge, knowledge is a specially secure kind of human belief, and scientific knowledge is a specially secure kind of human knowledge. Knowledge is subjective: it consists in justified, true belief, or in 'the right to be sure' (Ayer, 1956, p. 35). The assumption that knowledge necessarily involves truth has been dubbed the K-K thesis, and it is still asserted without question in standard studies in the theory of knowledge. If 'S knows that P' implies that 'P is true', and P is subsequently discovered to be false, then it follows that S did not know P after all. This is patently absurd, because if this truth condition is imposed, scientific knowledge is impossible. Popper's point is that we do have scientific knowledge.

Subjective knowledge, then, is largely dispositional: *my* knowledge consists of *my* dispositions, *your* knowledge consists of *your* dispositions — dispositions to act in certain ways, or to believe or say certain things. The conscious contents of the individual thinker set the bounds of the problem. Traditional epistemology concentrates on knowledge claims made by a person. It concerns itself with the question 'What is knowledge?' and engages in a search for the foundations of knowledge, for justification of knowledge claims.

The characteristics of this tradition are authority and certainty. Given that our knowledge can be traced back to an indubitable authority, no further justification is required, and the notion of certainty is engendered. Thus the traditional questions of epistemology are authoritarian in structure in the sense that they beg authoritarian answers. The two main streams of philosophy since Descartes and Locke, the rationalist (or intellectualist) and the empiricist, are tarred with the same brush in their attempts to establish certain, indubitable foundations on which to build the edifice of knowledge. The veracious God of Descartes or the sense experience of Locke would not deceive us.

Knowledge in the objective sense became possible with the development of the higher functions of language; it consists of spoken, or more importantly of written, statements. Thus the theories of natural science, once formulated and expressed and recorded linguistically, become objective. Newton's or Einstein's theories, as they developed in the minds of their creators, and as personal dispositions, are subjective; but once they become formulated in words and written down, they become objective. The crucial difference between objective and subjective knowledge is that between knowledge I have 'in my head' and knowledge put into words and written down. My subjective knowledge is part of myself, but as soon as it is vicariously represented in language, preferably in propositional form, it becomes objective and therefore criticisable. Scientific knowledge in particular, stored in books and libraries, is objective knowledge, and as such is available for scrutiny and criticism; for test and refutation.

On the pre-scientific level we hate the very idea that we may be mistaken. So we cling dogmatically to our conjectures, as long as possible. On the scientific level, we systematically search for our mistakes, for our errors. This is the great thing: we are consciously critical in order to detect our errors. Thus on the pre-scientific level, we are often ourselves destroyed, eliminated, with our false theories . . . On the scientific level, we systematically try to eliminate our false theories — we try to let our false theories die in our stead. This is the *critical method of error elimination*. It is the method of science. It presupposes that we can look at our theories critically — as something outside ourselves. They are not any longer our subjective beliefs — they are our objective conjectures. (1971, p. 73)

World 3. Thus the important feature of criticism, as Popper conceives it, is that it is objective; he rejects as uninteresting and misleading attempts to reduce knowledge to the psychology of the knower. Knowledge does not grow as a result of a study of human consciousness. In science we are interested in the logical status of theories, not in the psychology of its practitioners; or, more generally, in the argument, and not in the man, as the subject of our criticism. This led Popper to develop his theory of the three worlds in which he distinguishes between the objective world of material things, W1; the subjective world of individual minds, W2; and the world of objective structures and abstract ideas produced by minds, W3. Of particular significance in the history of World 3 is the emergence of language which gives rise to linguistically formulated objective ideas or theories, and thus to the emergence of criticism and therefore of rationality. World 3 theory has played an increasingly important part in Popper's later philosophy and, as we have observed, it is about the only one of the rich variety of Popper's ideas not contained in seminal form in his *Logic of Scientific Discovery*. It is invoked to explain the rationale of the three phases of evolution to which we have referred; physical, biological and cultural. More specifically it is used to explain Descartes' problem of the relationship between body and mind, and to solve Compton's Problem of the relationship between abstract meanings and human behaviour.[9]

Popper consequently breaks from a view of the world as a closed and determined system, and sees it as an open, indeterminate system which allows for growth, novelty, creativity and human freedom. Human freedom on the political level can only be achieved in an 'open society' which, in contrast to the 'closed society', allows for criticism and rejection of the prevailing government. Fundamental to Popper's conception of World 3 is that it is man-made and that therefore it contains theories that are both true and false; that it is open, or incomplete; and that although Worlds 1, 2 and 3 are partly autonomous, 'they belong to the same universe: they interact' (1982b, p. 161). Those criticisms of Popper's World 3 theory that liken it to a Platonic realm have not taken these points into consideration.

Fallibilism. The Popperian account of the growth of human knowledge through criticism rules out the notion of *episteme*, of sure and certain knowledge. The edifice of human knowledge is open and non-authoritarian in its structure. No ultimate justification for our knowledge claims is possible: the standard of rationality is set, not by an appeal to an indubitable authority, 'be it the Bible, the leader, the social class, the nation, the fortune-teller, the Word of God, the intellectual intuition, or sense-experience' (Bartley, 1964, p. 20), but by criticism. All human knowledge is fallible.

The abandonment of justificationary philosophy has an extraordinarily liberating effect. Once it is admitted that there is no possibility of ultimate justification in the realm of scientific knowledge, we need no longer embark on the fruitless and dangerous search for justification in the fields of politics or religion. What is of supreme importance is that we can criticise our theories in these fields.

> This is how we lift ourselves by our own bootstraps out of the morass of our ignorance; how we throw a rope into the air and then swarm up it − if it gets any purchase, however precarious, on any little twig. (1979, p. 148)

It is in this refusal on Popper's part to allow justification at any level whatsoever, that his critics see the fundamental weakness of his philosophy, which, they maintain, can only result in a sterile scepticism. To use Popper's analogy, the precarious little twig on which the rope gets purchase is an admission of justification, however provisional, and this, his critics insist, undermines his whole non-justificationary theory of knowledge.

Conclusion. Popper refers to evolution as a knowledge process; but the knowledge is largely subjective, it is part of the organism. The 'theories' of the organism are tested, and if falsified not only are the organism's theories killed, but the organism itself is eliminated because its theories are part of it.

With human knowledge the case is different because we formulate our theories linguistically in descriptive and argumentative language. Thus formulated, knowledge becomes objective, and as such can be objectively criticised. Our theories can be falsified, and if necessary rejected, but man, who has formulated the theories in his quest for understanding the world, does not perish with his false theories; they die in his stead.

The evolution of organisms and the evolution of human knowledge are represented by the critical feedback process of problem solving:

$$P_1 \rightarrow TS \rightarrow EE \rightarrow P_2$$

The evaluation is always critical, and its aim is the discovery and elimination of error. The growth of knowledge, resulting from error elimination, is a consequence of Darwinian selection, rather than Lamarckian instruction.

We must now spell out in greater detail the main themes contained in this summary account of Popper's epistemology, with increasing attention to the critical analysis to which they have been subjected.

Demarcation

Demarcation is not only 'the centre of the dispute', but it also provided the starting point for the growth of Popper's ideas. It was in Vienna, at the tender and impressionable age of seventeen, that the young Popper was confronted by conflicting claims and theories. In the field of physics there was Einstein who had followed his special theory of Relativity with the General theory of Relativity; in the field of psychoanalysis there were Freud and Adler; and on the political front there was Marx. There is no doubt that the theories associated with these names have had the most far-reaching consequences for twentieth-century man in his quest for understanding the mysteries of the universe, in his search for identity, and in his political struggles.

Popper's original insight was Darwinian in character. He detected that whereas Einstein's theory involved a risky prediction which could falsify it, the theories of Freud and Adler did not, and the risky predictions in Marxist theory were avoided by reformulating those parts of the theory that were in fact falsified by events – a process that became known as 'revisionism'. Thus Einstein's theory of General Relativity, in predicting that light would be bent in a strong gravitational field, suggested a means of testing the prediction. Eddington's famous eclipse experiment of 1919 showed that light was indeed deflected by the sun. Since that time the theory has successfully survived further tests, and has in fact proved most fruitful in modern cosmology. The theories of Freud and Adler and Marx, on the other hand, were not exposed to the risk of falsification but they were paraded as scientific theories on the strength of the confirming evidence that abounded.

> A Marxist could not open a newspaper without finding on every page confirming evidence for his interpretation of history; not only in the news, but also in its presentation — which revealed the class bias of the paper — and especially of course in what the paper did *not* say. The Freudian Analysts emphasised that their theories were constantly verified by their 'clinical observations'. As for Adler, I was much impressed by a personal experience. Once, in 1919, I reported to him a case which to me did not seem particularly Adlerian, but which he found no difficulty in analysing in terms of his theory of inferiority feelings, although he had not even seen the child. Slightly shocked, I asked him how he could be so sure. 'Because of my thousandfold experience,' he replied; whereupon I could not help saying: 'And with this new case, I suppose, your experience has become thousand-and-one-fold.' (1972, p. 35)

The fact that psychological and psychoanalytical theories appeared to be unfalsifiable did not necessarily render them useless, still less meaningless. They proved to some extent fruitful in providing an explanatory framework of human behaviour, but they are more in the nature of explanatory myths than testable scientific hypotheses. Such explanatory myths are not verified by the confirming evidence adduced by their adherents; to think otherwise leads to a sense of false security and an uncritical faith in the rightness of the theory. Popper does not deny that out of such pre-scientific myths testable scientific hypotheses might one day emerge, but this will only be when the theory in question makes a risky prediction that can in principle be falsified. The problem with Marxism is that once it frees itself from criticism, by not specifying in advance what would falsify it, it becomes a political slogan which enslaves rather than liberates its adherents. Much the same can be said for religious dogmas when isolated from all critical discussion.[10]

Thus Popper was led to propose what he later called his *criterion of demarcation* between science and non-science (but not between true and false theories): that 'the criterion of scientific status of a theory is its falsifiability, or refutability, or testability' (1972, p. 37). But from the beginning Popper realised that his proposal was problematic because, as we shall see, conclusive falsification proved to be an elusive goal. As early as 1934, in the *Logik der Forschung* he outlined the difficulties and he has spent a lifetime attempting to deal with them in a satisfactory manner.

However, the initial reaction to Popper's demarcation criterion, and one that persisted for a long time, was that he was proposing an alternative criterion of meaning to the Verification Criterion of the Logical Positivists:

> the fact that the positivists used their criterion of verifiability and meaningfulness also as a criterion of demarcation made them deaf and blind to the fact that I used falsifiability as a criterion of demarcation, but never of meaningfulness. (1974, p. 967)

This misunderstanding highlights one of the fundamental differences between Popper and the positivists, at the same time implying far-reaching consequences for metaphysical conjectures (including those of religion). The positivists' attempt to kill metaphysics by legislating between sense and nonsense leads, as we have observed, to the paradoxical consequence that it rules out science itself. Popper's criterion of demarcation, on the other hand, proved to have the most fruitful consequences because 'it prepares the way for a theory of testability and content, and for a solution of the problem of induction' (1974, p. 964).

Initially Popper treated demarcation and induction as separate problems, but he saw how closely they are linked when he realised that the positivists had used induction as the distinguishing criterion of demarcation between science and metaphysics. Popper's demarcation proposal coincided with his solution to the problems associated with induction: with falsification as the criterion of demarcation, Popper could discard induction without getting into trouble over demarcation. This was the first of many links that Popper made between the variety of problems that eventually merged into his evolutionary epistemology. As we have noted, the original demarcation proposal itself turned out to be part of the wider and more general philosophical problem of demarcating between criticisable and non-criticisable theories. There is no doubt that Popper has been foremost in the post-positivist rehabilitation of metaphysics.

We must return to Popper's original attempt to demarcate between science and non-science. If Popper cannot show that knowledge grows though criticism, then he must concede that one cannot talk about the growth of knowledge as a rational process. This indeed is at the heart of the challenge that Kuhn made in *The Structure of Scientific Revolutions*, which we will examine, along with Feyerabend's development of it, in our next chapter. Falsification is the key to rational criticism in science, and it is Popper's contention that it is only through falsification that we can approach the truth, which is the aim of science. It is at these points that Popper's philosophy of science is exposed to the most searching and serious criticism, even to the point of Feyerabend's denial that there can be *any* demarcation between rationality and irrationality, let alone between science and non-science. We must be clear that what is at stake here is the fundamental question of rationality in the affairs of men.

In his reply to his critics in the Schilpp volume, Popper attempts to deal with the problems associated with demarcation, and we will examine these in conjunction with his treatment of induction. But before we turn to induction it is instructive to note Popper's preface to his reply to the criticisms of his demarcation proposal. He admits that although the objects of our study, whether swans, or stars, or philosophers, are complex, we should aim at simplifications which highlight the main idea. 'To put it briefly: never lose sight of the main idea. Never get involved in little side issues if they can be avoided or solved in a simple and straightforward way' (1974, p. 977). He then admits that there is a sense in which his view of science is selective, and he names such great pioneers as Galileo, Kepler, Newton, Einstein and Bohr as representative of science in the heroic sense that interests him. 'They work with bold conjectures and severe attempts at refuting their own conjectures,' and he continues, 'My criterion of demarcation between science and non-science is a simple and logical analysis of this picture. How good or bad it is will be shown by its fertility' (1974, p. 978).

The characteristic of science in this sense is the manner in which its theories are exposed to the risk of being false. Thus Copernicus made the daring conjecture that the sun, rather than the earth, rests at the centre of the universe. It turned out to be false, but immensely fertile in that it led, through Galileo, to Kepler's heartrending struggle to establish his three laws of planetary motion, which in turn proved to be but approximations to the truth. It took the genius of Newton to develop the ideas into the most impressive physico-mathematical theory that had ever been conceived, only to be falsified by Einstein's even more impressive theory of gravitation. Acquaintance with Einstein's theory was, for Popper, a converting experience that not only led him into the philosophy of science, but which remained for him the paradigm of all great science. Indeed, one of the criticisms levelled at

Popper is that in concentrating on the Einsteinian 'heroic' example, he ignores the great variety and diversity that the history of science exhibits. This is why Kuhn, as a historian of science, challenged Popper in the manner that he did.

In his 1982 introduction to *Realism and the Aim of Science* Popper attempts to set the record straight with regard to his methodology and the history of science, with Kuhn's work particularly in mind. He emphasises that although his methodology is a philosophical or metaphysical discipline, more in the nature of a normative proposal, it is of great interest for the historian of science. Whilst putting in a disclaimer as a historian of science, he lists 'at random' some twenty examples in the history of physics in which refutations led to revolutionary theoretical reconstructions, from early Greek to twentieth-century atomic theory. He concludes:

> My theory of science was not intended to be an historical theory, or to be a theory supported by historical or other empirical facts . . . *Yet I doubt whether there exists any theory of science which can throw so much light on the history of science as the theory of refutation followed by revolutionary and yet conservative reconstruction.* (1983, p. xxxi)

However, the fact is that Popper's philosophy of science in general, and his criterion of demarcation in particular, is derived from the Einsteinian example of postulating a bold theory which can in principle be refuted, and which, even if not falsified, is still only an approximation to the truth. And that, of course, is why Einstein devoted the latter half of his life working unsuccessfully on a Unified Field Theory which would contain General Relativity as an approximation.

Thus Popper pictures the great scientist as one who makes bold and daring conjectures which attempt to go beneath the world of appearances and penetrate into the deeper realities with his explanatory theories. It is at the level of appearances, at the level of our observations, that these bold theories take the risk of being tested and refuted; of clashing with reality. And this is the principle behind Popper's demarcation proposal. Empirical science is distinguished from non-science, and from pre-scientific myths and metaphysics by its boldness and the attendant risks of refutations.

Quite clearly Popper's demarcation criterion is not itself open to empirical falsification because it is a methodological proposal which is offered as a loose definition of science. Anyone is free to accept or reject it, or to propose an alternative definition of science. Popper's approach is to draw our attention to the Einsteinian example and suggest that this should set the standard for the growth of scientific knowledge. The proposal is not so much a moral imperative as a hypothetical imperative which belongs to the logic of the situation. Popper admits, under pressure from Kuhn, that scientists have not always adopted the Einsteinian critical methodology of bold conjectures and severe testing, but he insists that methodological avoidance of criticism leads to stagnation rather than growth (1970).

Nevertheless, as a philosophical theory (of science) Popper's methodology is, by his own standards, open to criticism and rejection. He accepts a challenge of Lakatos by saying that he is prepared to abandon his demarcation criterion if it can be shown that 'Newton's theory is no more falsifiable by "observable states of affairs" than is Freud's' (1974, p. 1010). The significance of this challenge will become clearer as we discuss Popper's methodology in relation to his treatment of induction.

The Myth of Induction

Popper's rejection of induction as a means of acquiring scientific knowledge is one of the central themes running through his whole philosophy. His excursion into the biological realm and his attachment to the Darwinian natural selection idea, which developed into an evolutionary epistemology, was a direct consequence of his treatment of the age-old problems associated with induction. There is no doubt that Popper's treatment of induction has stimulated the most fruitful critical discussion of some of the fundamental assumptions in which our thinking and acting are rooted. Embedded in this discussion are the perennial problems of knowledge, truth and realism that have bewitched and bemused philosophers since Plato. Bertrand Russell, the twentieth century's great advocate of rationalism, was certainly bemused by the problem of induction when he asserted that if it could not be satisfactorily resolved there could be 'no intellectual difference between sanity and insanity', and that therefore we would find it difficult to counter the lunatic who claims he is a poached egg (Russell, 1946, p. 699).

In addressing himself to the problem of how we get knowledge from the world, Popper distinguishes between what he calls common-sense realism and the common-sense theory of knowledge. All our knowledge begins with common-sense views, but it is only through criticism of these assumptions that we discover their worth. He is a strong advocate of common-sense realism: that is the view that there is a real world, full of interesting things, which exists independently of our ideas and perceptions of it. Most people, and not least practising scientists, do not question this view; and even those philosophers who tend to idealism, such as Berkeley, Hume and Mach, who at the intellectual level cannot escape from the imprisonment of their minds, behave like the rest of us who accept the objective existence of real people and sticks and stones. Although neither realism nor idealism can be conclusively established, Popper maintains that the arguments for realism are far stronger than those against it, and that no valid criticism of realism has ever been proposed. He argues this thesis in many places, at one point even calling Winston Churchill to his aid as a hitherto unrecognised epistemologist (1979, pp. 42-3), and most comprehensively in *Realism and the Aim of Science* (1983). He takes as his standard criterion of reality the example of Dr Johnson, who demonstrated that if you kick a stone it will 'kick back' (1982b, p. 116).

The theory of knowledge is largely concerned with how we get knowledge of this world and ourselves as part of it. The traditional, common-sense view is: by opening our eyes and ears; that is, through our senses. Popper has dubbed this as the *bucket theory of the mind* because it depicts the mind as a sort of receptacle in which knowledge that enters through the senses accumulates (1979, Appendix 1). This common-sense theory of knowledge, which is characteristic of British empiricism, leads to a kind of anti-realism because it implies that the world consists of the contents of our minds; that is, ideas resulting from sense impressions. This was precisely the dilemma that Hume and his positivist successors found themselves in. The real world is reduced to the contents of our minds. Thus the naive empiricist is led through idealism to its logical conclusion, solipsism: the doctrine that only a philosopher could entertain, that the world is my dream. The absurdity of this view is exposed by the story of Lady Ottoline Morrell who, it is alleged, wrote to Bertrand Russell confessing that she was a solipsist and could not understand why she could persuade nobody to join her!

Popper rejects this common-sense, inductive theory of knowledge, despite its impressive historical pedigree, and he replaces it with an account of the growth of knowledge which is consistent with the descriptive epistemology we have so far outlined. His thesis, that knowledge is conjectural rather than imposed on us from without, arose out of his consideration of Hume's problem regarding induction. The mind, instead of being a passive receptor, or 'bucket', is more like a *searchlight* which plays an active part in the process of perception. Following Hume, Popper divides the problem of induction into two: one logical, the other psychological. In *The Logic of Scientific Discovery* he wrote:

> The fundamental doctrine which underlies all theories of induction is *the doctrine of the primacy of repetitions*. Keeping Hume's attitude in mind, we may distinguish between two variants of this doctrine. The first (which Hume criticized) may be called the doctrine of the logical primacy of repetitions. According to this doctrine, repeated instances furnish a kind of *justification* for the acceptance of a universal law. (The idea of repetition is linked, as a rule, with that of probability.) The second (which Hume upheld) may be called the doctrine of the temporal (and psychological) primacy of repetitions. According to this second doctrine, repetitions, even though they should fail to furnish any kind of *justification* for a universal law and for the expectations and beliefs which it entails nevertheless induce and *arouse* these expectations and beliefs in us, as a matter of fact — however little 'justified' or 'rational' this fact (or these beliefs) may be. (1968a, p. 420)

Popper accepts Hume's rejection of the logical primacy of repetitions, and his conclusion that induction is irrational; but he rejects the psychological primacy of repetitions, maintaining that such a clash between logic and psychology is unacceptable (1974, p. 1020).

If, as Popper stoutly maintains, science is a rational activity, and induction is irrational, then induction must be a myth, an intellectual fraud, a 'kind of optical illusion' (1974, p. 1015). Not only do we not need induction, but we do not in fact use it. How then can we account for our expectations of future events which appear to be common-sense inductive inferences, and how can we account for the rational nature of science? Let us take these questions in turn.

Induction — the Psychology of the Situation

The Psychological Primacy of Repetitions. Although Hume was forced by his logic to reject induction as a rational inference, he nevertheless retained it as psychologically unavoidable and biologically indispensable. In fact for Hume inductive inference is as natural and necessary as breathing. He relates in his *Enquiry* how even the hardened philosophical sceptic must proceed in this irrational manner (Part II, Section 12) and we have noted how his positivist successors adopted the same conclusion. However, Popper's 'principle of transference', that what is true in logic must be true in psychology (1974, p. 1020) leads him to reject Hume's psychological account of induction on logical grounds.

He demonstrates that Hume's thesis, that constant repetition leads to expectations, leads to an infinite regress, because a single repetition must be recognised as a repetition. Similarity is always similarity in a certain respect and from a point of view which is adopted as a result of a choice or decision.

In rejecting this psychological thesis, Popper turns the tables on Hume, and instead of explaining our propensity to expect regularities as a result of repetition, he proposes to explain repetition-for-us as a result of our propensity to expect regularities and search for them. We only recognise regularities because we have a biological, in-built propensity to do so. Contrary to Hume, then, Popper's evolutionism maintains that it is the method of trial and error that is as natural as breathing.

> Thus I was led by purely logical considerations to replace the psychological theory of induction by the following view. Without waiting, passively, for repetitions to impress or impose regularities upon us, we actively try to impose regularities on the world. We try to discover similarities in it, and to interpret it in terms of laws invented by us. Without waiting for premises we jump to conclusions. These may have to be discarded later, should observation show they are wrong.
>
> This was a theory of trial and error — of *conjectures and refutations*. It made it possible to understand why our attempts to force interpretations on the world were logically prior to the observations of similarities. Since there were logical reasons behind this procedure, I thought that it would apply in the field of science also; that scientific theories were not the digest of observations, but that they were inventions — conjectures boldly put forward for trial, to be eliminated if they clashed with observations; with observations which were rarely accidental but as a rule undertaken with the definite intention of testing a theory by obtaining, if possible, a decisive refutation. (1972, p. 46)

Although the observation of repetitions, central to the inductive method, cannot be achieved in the presuppositionless way required by induction, it cannot be conclusively established that the observation of repetitions plays no part at all in the formation of theories. Nevertheless Popper has succeeded in giving a far more plausible account of theory formation by his emphasis on the formulation of bold, imaginative hypotheses, and one which is nearer the truth than the view of the scientist as a collector and collator of data. The view that scientific theories are a product of the creative imagination is supported by Einstein (1961, p. 294),[11] and by accounts such as that given by Watson in *The Double Helix* (1968).

The creative scientist, then, is motivated by his theoretical interests, rather than by neutral observations. Popper continues;

> The problem 'Which comes first, the hypothesis (H) or the observation (O),' is soluble; as is the problem, 'Which comes first, the hen (H) or the egg (O)'. The reply to the latter is, 'An earlier kind of egg'; to the former, 'An earlier kind of hypothesis'. It is quite true that any particular hypothesis we choose will have been preceded by observations — the observations, for example, which it is designed to explain. But these observations, in their turn, presupposed the adoption of a frame of reference: a frame of theories: a frame of expectations. If they were significant, if they created a need for explanation and thus gave rise to the invention of a hypothesis, it was because they could not be explained within the old theoretical framework, the old horizon of expectations. There is no danger here of an infinite regress. Going back to more and more primitive theories and myths we shall in the end find unconscious, *inborn* expectations. (1972, p. 47)

Inborn Expectations. This theory of inborn expectations provides a direct link between the evolution of knowledge and biological evolution. A newborn animal does not patiently wait to have beliefs instilled into it by constant repetition of sense experiences; rather, it instinctively relates to the environment. This 'instinctive' behaviour is the result of a long evolutionary process of trial and error during which the organism has developed a hierarchical system of controls which enables it to cope with the problems that normally confront it. The system will tend to break down if the normally stable conditions radically change, but providing stable environmental conditions exist the animal may be said to be born with expectations or foreknowledge. A newborn baby 'expects' to be fed, it 'knows' what to do when presented with the nipple; we could say that it 'expects' to be loved and cared for. Locke could not have been further from the truth in this matter in his *tabula rasa* theory of learning and acquiring knowledge. Hence the relevance of an adequate descriptive epistemology.

The expectation of finding regularities is psychologically or genetically, and logically *a priori*, but it is not valid *a priori*, for it may fail. We could construct a chaotic environment in which we fail to find regularities, and such an environment could well be lethal. One could account for schizophrenia in these terms; the illness resulting from an inability (due to a genetic or epigenetic failure) to find the needed regularities in the world.

Sense Experience. What, then, is the relationship between inborn expectations and sense experience? If we reject the empiricist account of inductive knowledge arriving through the senses, what part does our sensory equipment play? On the evolutionist view our senses do not present us directly with raw data for our minds to process; rather, our sensory equipment presents us with *theories*, with information we need for problem solving. The point here is that the process of interpretation does not begin in the mind, but at the physiological level in the sense organs themselves. On this view there is no such thing as a pure, neutral sense datum, or even the logical 'qualia' preferred by Ayer (1976, pp. 71-2). Although we cannot inhibit our in-built interpretive propensities, our sensory equipment is by no means infallible; it can make mistakes. The visual system, for instance, is far from perfect. As Campbell puts it, 'We are misled by double images, blind spots, optical illusions, chromatic aberrations, astigmatism, venous shadows etc.' (1974, p. 414). Hence there can be no satisfactory epistemology assuming veridical visual perception. This is consistent with the logic of Popper's fallibilism which will not admit justification even at the level of our most basic observations. Nevertheless the in-built propensity for finding regularities implies that our sense organs are by and large pretty good at decoding signals reaching them from the environment. If our sense organs were chaotically unreliable we would not be here, for, as we have observed, the evolutionary natural-selection elimination process has enabled us to organise and understand our environment. If this were not so we would long ago have been eliminated.

Induction – the Logic of the Situation

Logical Falsification. Popper follows Hume in accepting deduction as the organon of logic. He does not attempt to justify his attachment to deductivism intuitively, for by the same token induction could be 'justified'. This recourse to self-evident truth has been described as the 'analytic justification', and, as we have indicated, it has been courted by many philosophers troubled by Hume's discovery regarding the logical primacy of repetitions. However, Popper maintains that a deductive argument can be judged by objective standards which take it out of the realm of subjective belief. A deductive argument is valid if truth is transferred from the premiss to the conclusion; that is, if no counter-example exists (1976, p. 146). No such rule for *inductive* inference has ever been proposed.[12] Inductivists can only produce such vague and uninteresting rules as 'the future is likely to be not so very different from the past', which is no rule at all because it allows the future to be rather different from the past, which it often is. The rationality of induction certainly cannot be established by showing that it is reliable. We have mentioned Russell's unfortunate chicken; Popper offers more serious examples of failures of inductively based expectations, such as the midnight sun, and cases of ergotism, which count against the inductively formed beliefs that the sun rises and sets every twenty-four hours, or that bread nourishes (1979, pp. 10-11).

Popper does not deny the importance of subjective, metaphysical beliefs concerning regularities and expectations. In *The Logic of Scientific Discovery* he refers to 'the metaphysical faith in the existence of regularities in our world (a faith which I share, and without which practical action is hardly conceivable)' (1968a, p. 252). Nevertheless 'for rational action we need criticism even more urgently than faith' (1974, p. 1041).

If, then, induction is out on logical grounds, what is the characteristic of science that renders it a rational affair? Popper's answer is to turn the tables once again on Hume and the inductivists by demonstrating that science is in fact deductive, and by that token rational. He arrives at this conclusion by reformulating Hume's problem in terms of statements rather than 'instances' of which we have had experience, and he concludes that although a universal theory cannot be justified it can sometimes be falsified by *'the assumption of the truth of test statements'* (1979, p. 7). In other words, *if a counterexample exists* the theory is falsified. It is important to see what Popper has done here because many commentators have not recognised the distinction between the *logic* of the argument and its pragmatic implications. *As a matter of logic* Popper is surely right; given the truth of a test statement, which is a counterexample, a theory is conclusively falsified.

> In other words, from a purely logical point of view, the acceptance of *one* counterinstance to 'All swans are white' implies the falsity of the law 'All swans are white' — that law, that is, whose counterinstance we accepted. Induction is logically invalid; but refutation or falsification is a logically valid way of arguing from a single counterinstance to — or, rather, against — the corresponding law.
>
> This logical situation is completely independent of any question

of whether we would, in practice, accept a single counterinstance —
for example, a solitary black swan — in refutation of a so far highly
successful law. (1974, pp. 1020-1)

Popper's and Hume's negative result establishes that all our theories
remain forever guesses, hypotheses, conjectures, but it can also be used
to support a positive theory in terms of our preferences in the arena of
competing theories. The preference is guided by the idea of truth and
greater explanatory power. Our only way of approaching truth is by
eliminating falsehood. This is a sufficient answer to Russell's problem
regarding the difference between the scientist and the lunatic. Both the
scientist's and the lunatic's theories are conjectural, but maybe the
lunatic's theory can be refuted by observation, which is a rational procedure.

So much for the logic of the situation. Rather than describe this, as
Kuhn and Lakatos and subsequent critics have misleadingly done, as
naive falsification, it would be more appropriately labelled as *logical*
falsification, to distinguish it from the more sophisticated and problematic falsification that has to be adopted in practice.[13]

Methodological Falsification. In our examination of positivism we
observed how the attempts to apply the logic of the *Principia* to the
theories of the scientists came to grief because of the complexity of
scientific theories. Popper, who, as we have noted, is far more sensitive
to the actual practice of scientists, nevertheless makes a similar attempt
at applying the rigours of deductive logic to scientific theorising. His
demarcation between science and pseudo-science is founded on a formal
logical principle; that of logical falsification, which is an epistemological
version of *modus tollens*. And yet it is through an examination of the
content of scientific theories that the logic of the situation is put under
serious, and some would say fatal, stress. The problem is that what is
true in logic is often not true in life. Many a great philosopher, from
Plato onwards, has foundered on this paradox. Popper, nevertheless,
makes a valiant attempt at defending his formal, logical principle of
demarcation by falsification whilst attempting to take account of the
realities of scientific practice.

The problem Popper was confronted with right from the beginning
was that although the logic of falsification appeared to be impeccable,
in practice no decisive falsification can ever be achieved. There are a
number of reasons for this, the fundamental one being what is called
in *The Logic of Scientific Discovery* 'the problem of the empirical basis'

(Ch. 5). This problem derives from the fact that none of our observations can be ultimately justified because all our observations are 'theory soaked'. The flaw in the foundations of the empirical tradition results from a failure to recognise that observation cannot be prior to theory, and that is why the efforts of the Logical Positivists to eliminate the theoretical components from theories came to grief. Popper recognised this when he suggested in his *Logic of Scientific Discovery* that even the statement 'Here is a glass of water' cannot be verified by observation because 'glass' and 'water' are universal terms and imply law-like behaviour (1968a, pp. 94-5). This early insight of Popper's was later complemented by his biological account of sense experience that we have outlined. These ideas regarding the theoretical nature of perception and the denial of a neutral observation language were later taken up by other philosophers of science, notably Hanson, Feyerabend and Kuhn, who exploited the original Popperian insight to support the most un-Popperian theses of relativism and epistemological anarchism.

If even the most basic observation statement has a theoretical component, and therefore cannot be justified, how can we set about attempting to falsify the general theories of science by observations? Popper's way out of this dilemma is to confer a special role to certain 'basic' statements which are singular, existential statements about some definite spatio-temporal region, such as 'There is a raven in spatio-temporal region k.' The role of such statements is not to provide an epistemological bedrock, for, 'our "basic statements" are anything but "basic" in the sense "final"; they are "basic" only in the sense that they belong to that class of statements which are used in testing our theories' (1972, p. 388). Hence, basic statements are not special types of ultimate, or protocol, statements but statements accepted for testing theories. That is why Popper now prefers to call them 'test statements'. The acceptance of these test statements depends upon the decisions of the scientific community, which Popper likens to the agreement of a jury. 'From a logical point of view, the testing of a theory depends upon basic statements whose acceptance or rejection, in its turn, depends upon our *decisions*. Thus it is *decisions* which settle the fate of theories' (1968a, p. 108).

The admission of this subjective, community element in scientific practice is a far remove from the positivist notion of science founded on what is objectively and indubitably given. Not only does it throw up interesting implications for religious thinking, but it opens up the whole question of science as a social enterprise, which is the subject of our next chapter.

Popper sums up his view in what has become a classic statement of fallibilism:

> The empirical basis of objective science has thus nothing 'absolute' about it. Science does not rest upon solid bedrock. The bold structure of its theories rises, as it were, above a swamp. It is like a building erected on piles. The piles are driven down from above into the swamp, but not down to any natural or 'given' base; and if we stop driving the piles deeper, it is not because we have reached firm ground. We simply stop when we are satisfied that the piles are firm enough to carry the structure, at least for the time being. (1968a, p. 111).

Although most philosophers agree with the Kantian notion, developed by Popper, that the perceiver contributes to what he perceives, and that therefore observation and theory are interdependent, there are many who cannot accept Popper's unrelenting fallibilism and unwillingness to admit justification at any level, however provisional. O'Hear, for example, using Wittgenstein's private language argument, maintains that a cohesive conceptual scheme needs to have some grounding in talk about enduring objects in the world. This need for agreement in language is not consistent with a constant questioning of our basic assumptions. He maintains, 'doubting clear-cut everyday judgements about enduring objects would be reasonable only given a general breakdown of world order or perceptual order in which we would also lose the ability to categorise even our perceptions' (O'Hear, 1980, p. 87). Popper has never denied this; indeed his biological account of sense experience in terms of inborn expectations takes full account of O'Hear's point. Popper's contention is that if our knowledge is to grow we must *consciously* adopt a critical attitude, and this is why he will not admit justification, even at the deepest level.

However, even if agreement is reached about the acceptance of a test statement as a 'potential falsifier', conclusive falsification is still not possible, nor is it always desirable, as the history of science demonstrates. There are many examples of the falsification of theories, which, had they been accepted, would not have led to the growth of scientific knowledge. Consequently Popper has had to loosen considerably the application of his criterion of demarcation by falsification.

Although philosophers of science talk about white swans and black ravens, the theories of science are in fact more complex structures. It is therefore difficult, if not impossible, to say which part of the complex system of interrelated statements that constitutes a theory has been exposed to a particular experimental test. Popper has from the beginning been aware of this problem which arises from the logic of *modus tollens*, according to which the falsification of a conclusion entails the falsification of the system from which it has been derived (1968a, Section 18). Quine supports this conclusion when he says that '" All ravens are black" is the very paradigm, in miniature, of an empirical law' (Quine, 1974, p. 220). Consequently, maintains Popper, 'Newton's theory is a system. *If we falsify it, we falsify the whole system*' (1974, p. 982). Further conjectures, or hypotheses, are needed to attempt to establish which particular part of the system is at fault, and these are in turn tested.

Popper also agrees that evasive tactics, or 'immunizing stratagems', can be adopted in the face of refutations, but he distinguishes between *auxiliary* and *ad hoc* hypotheses. An *auxiliary* hypothesis, introduced to explain a particular difficulty, can itself be tested, whereas an *ad hoc* hypothesis is one that it is not possible to test independently. A classic example from astronomy was the defence of Newton's theory by introducing the auxiliary hypothesis of an undetected planet to account for observed perturbations in the orbit of Uranus. Thus the planet Neptune was detected by optical means and 'the threatened refutation of Newtonian theory was transformed into a resounding success' (1974, p. 986). The case of the neutrino in the theory of matter is rather more complex. Pauli introduced the *ad hoc* hypothesis of the neutrino, an extraordinarily elusive particle, containing no mass, which appeared at the time to be in principle undetectable; but with the growth of scientific knowledge shed its *ad hoc* character and was successfully identified. Lakatos confuses the situation even more by introducing a hypothetical example of a theory which could always be protected by the introduction of endless auxiliary and *ad hoc* hypotheses (Lakatos, 1970, pp. 100-1). For these reasons John Krige, in his *Science, Revolution and Discontinuity* (1980), reckons that Popper's falsification methodology, far from lighting the fuse of scientific revolution, is conservative with regard to new research programmes.

Popper resolutely maintains that all these difficulties can be overcome provided we adopt the mean between falsification at any price (Lakatos' example) and rejecting the theory without debate. Consequently the formal, logical principle of demarcation takes the methodological form: 'Propose theories which can be criticized. Think about possible decisive falsifying experiments — crucial experiments. But do not give up your theories too easily — not, at any rate, before you have critically examined your criticism' (1974, p. 984). And in the 1983 introduction to *Realism and the Aim of Science* he adds that 'the uncertainty of every empirical falsification . . . should not be taken too seriously', and he cites Rutherford's refutation of the Thomson model of the atom, which led Rutherford to propose the nuclear model, as an illustration of the force which a falsification may have. In this area of nuclear physics we have examples of falsifications following fairly fast, one after the other; the Rutherford atom being replaced by the Bohr model, which it turn was superseded by quantum mechanics (1983, pp. xxxiii-xxxiv).

Thus what looked initially like a clear, formal principle of demarcation by falsification turns out to be in practice somewhat loose; and this, Popper admits, is because 'the transition between metaphysics and science is not a sharp one: what was a metaphysical idea yesterday can become a testable scientific theory tomorrow' (1974, p. 981). The fact that Popper is led to such extremes to distinguish between the refutation of a theory and its rejection in practice illustrates the tension between his formal, logical approach and his sensitivity to the realities of scientific practice. The tension is never resolved, nor can it be. For that reason some philosophers, whom we will discuss in the next chapter, have abandoned the formal, logical approach for a view of science which depicts it as an essentially irrational enterprise which, at best, in all its vagaries, can only be described.

Truth and Verisimilitude

Tarski and the Correspondence Theory of Truth. Essential to the logic of the situation is the idea of truth. If criticism is the method, truth is the aim. On the face of it, this sounds perverse in view of Popper's denunciation of the inductivist search for true theories and his insistence that verification must be replaced by falsification. If we can never establish the truth, not just of scientific theories, but of even the most humble 'basic' statement resulting from observation, how can we ever begin to talk about truth? This was indeed Popper's position when he wrote *Logik der Forschung* and it was not until he encountered Tarski's idea of truth that he was prepared to offer a rational account of truth, rather than simply admit it as a metaphysical belief, or a pious hope. It was Tarski's rehabilitation of the correspondence theory of truth that allowed Popper to become far more positive in his approach and to spell out the aim of science in terms of true explanatory theories (1979, Ch. 5).

The notion of truth as an absolute and objective goal is fundamental to the whole of Popper's thinking, and it is this that gives his realism the necessary philosophical backing and sets him clearly against the strong currents of relativism and scepticism. He wrote in 1961:

> The main philosophical malady of our time is an intellectual and moral relativism, the latter being at least in part based upon the former. By relativism − or if you like, scepticism − I mean here, briefly, the theory that the choice between competing theories is arbitrary; since either, there is no such thing as objective truth; or, if there is, no such thing as a theory which is true or at any rate (though perhaps not true) nearer to the truth than another theory; or, if there are two or more theories, no ways or means of deciding whether one of them is better than another. (1966, Vol. 2, p. 369)

The problems of relativism, particularly the relationship between our conceptual systems and the real world, will be pursued in the next chapter, but clearly, a logically consistent notion of truth is necessary for an objective and realistic theory of science such as Popper's. It is also essential if we are to distinguish between pure and applied science; that is, to make a distinction between the search for knowledge and the search for power, a distinction that is dangerously blurred by the success of our technology.

> For the difference is that, in the search for knowledge, we are out to find true theories, or at least theories which are nearer than others to the truth − which correspond better to the facts; whereas in the search for powerful instruments we are, in many cases, quite well served by theories which are known to be false. (Popper, 1972, p. 226)

Although the correspondence theory of truth is intuitively the most appealing, attempts to spell out the correspondence between a statement and a fact, that is, between language and the world, have proved strangely elusive, as we have indicated in our discussion of Wittgenstein's attempt to clarify the relationship in terms of his picture theory. However, all was beautifully clarified for Popper when Tarski took him aside and explained his theory of truth in a public park in Vienna in 1935.

What Tarski did was to give logical form to the intuitive idea that truth is correspondence to the facts. Although there is a degree of complexity in the technicalities of the logic involved, the result, as Popper

frequently emphasises, appears to be trivial and obvious; but its significance is profound, and for Popper remains one of the three most important ideas in the whole of his philosophy (the other two being
demarcation and indeterminism) (1974, p. 976).

Popper suggests that Tarski's result is more readily appreciated if we
assume truth to be correspondence to the facts and then proceed to explain the idea of 'correspondence to the facts' (1972, Ch. 10; 1979; Ch. 9).

Tarski's insight consists in distinguishing between the *semantical*
metalanguage in which we can speak about two things – (1) statements
and (2) the facts to which they refer; and the *syntactical* metalanguage
in which we can speak about the object language but *not* about the facts
to which it refers. In other words we need to distinguish between statements about statements, and statements about facts. To take a simple
example which exhibits this distinction: '*Snow is white' corresponds to
the facts if, and only if, snow is white*. This simple and obvious formulation, which highlights the semantic-syntax distinction, contains the
solution to the truth paradox which defeated Russell and Whitehead in
Principia Mathematica. Tarski offers a *definition* of truth in formal and
logical terms, and it allows us to speak of truth in an objective sense
rather than in terms of subjective belief.

Although some philosophers question whether Tarski's result can be
extended beyond formalised or 'artificial' languages, Popper insists that
the problem of applying the theory to 'natural' languages can be readily
overcome, as a result of which we may speak of theories as being absol-
utely true or false, and so view them realistically. Once again we must
point to the important distinction between the logic and the method,
which is characteristic of Popper's approach. Tarski's theory is a logical
theory, and Popper embraces it because the logical status of truth is
essential for his theory of rationality: that knowledge grows through
criticism; and growth, in this context, implies growth towards the truth.

Tarski has offered a definition of truth, but *not* a *criterion* by which
we may discover the truth of a particular theory. Such a criterion,
Popper points out, would render us omniscient. He reckons that 'it is
the demand for a *criterion of truth* which has made so many people feel
the question "What is truth?" is unanswerable'. And he continues:

> But the absence of a criterion of truth does not render the notion of
> truth non-significant any more than the absence of a criterion of
> health renders the notion of health non-significant. A sick man may
> seek health even though he has no criterion for it. An erring man
> may seek truth even though he has no criterion for it. (1966, Vol. 2

p. 373)

On the face of it, the characterisation of science as a rational activity, the goal of which is truth, is somewhat otiose if that goal is unrecognisable. But Popper maintains that we can be guided by the idea of truth as a regulative principle, and he offers the analogy of a mountain peak usually wrapped in clouds.

> A climber may not merely have difficulties in getting there — he may not know when he gets there, because he may be unable to distinguish, in the clouds, between the main peak and a subsidiary peak. Yet this does not affect the objective existence of the summit; and if a climber tells us 'I doubt whether I reached the actual summit,' then he does, by implication, recognise the objective existence of the summit. The very idea of error, or doubt (in its normal straightforward sense) implies the idea of objective truth which we may fail to reach. (1972, p. 226)

The last point is important. Truth and falsity are complementary terms, each depending on the other for its significance. We may recall Popper's biological suggestion about the idea of truth and falsity arising with the descriptive function of language. The scientist looks for theories that are true by subjecting them to severe tests in order to attempt to eliminate falsehood. Thus falsifiability ranks with truth as the aim of science.

Verisimilitude. Popper maintains that although we never know if and when we have reached the truth, we do have a criterion of progress: that is, we know when one theory is replaced by a better theory. The better theory tells us more about the world; that is, it has greater informative content, and therefore it can be more severely tested. This has the curious consequence that as empirical content increases, so probability decreases (in the sense of the probability calculus) and vice versa. In other words, content increases with increasing *im*probability; the better the theory, the more improbable it is. The positivists, who had substituted high probability for conclusive verification, were reluctant to concede that the growth of scientific knowledge implies low probability, and that therefore truth implies vanishing probability. For example, 'It will rain' is highly probable, but highly uninformative and also unfalsifiable; whereas 'It will rain at Old Trafford on Saturday afternoon' is less probable, but highly informative and easily testable

(1968a, Ch. 6, and 1972, Ch. 10).

Thus we look for bolder, more informative theories, remembering that the more a theory forbids, the more it tells us. We prefer Newton's theory to Kepler's, and Einstein's to Newton's because we have accepted counterexamples that have refuted the earlier theories. Popper sums this up in the form of a methodological rule: 'try out, and aim at, bold theories, with great informative content; and let these bold theories compete, by discussing them critically and testing them severely' (1974, p. 1022). This led Popper to combine the ideas of truth and content into his thesis of verisimilitude, or approach to the truth. To the extent that science consists of the practical business of producing true explanatory theories, the scientist must have some indication that at least he is approaching the truth with his theories. The metaphysical goal of truth must be earthed at some points in order that he knows if he is on the right track, otherwise his search would be aimless and futile.

Popper's attempt to establish a formal account of verisimilitude derives from his earlier account of corroboration. Using the Darwinian metaphor, a well-corroborated theory is one that has demonstrated its fitness by surviving past tests. But just as biological survival to date does not imply future survival, so with scientific theories. Popper forbids any inductive link from past performance to future expectations.

> What I call the corroboration of a hypothesis is merely a summary report about the past performance of the hypothesis; it is *not* an attempt to justify any expectation that the hypotheses will *in future* prove successful if it was successful in the past. Rather, I hold with Hume that *nothing* can justify such an expectation. (1974, p. 1043)

Many critics have therefore declared Popperian corroboration redundant on the grounds that past reports are only of interest if they have future implications. A school report based on past performance and examination results is only of use to a prospective employer if it gives some indication of future possibilities. Nevertheless, Popper insists that the well-corroborated theory is the *best* theory, not because of its implications for future reliability, but because there is nothing more rational than to choose the best criticised theory.

Popper makes the link between corroboration and verisimilitude, not by taking corroboration as a *measure* of verisimilitude, but simply as an *indicator* of verisimilitude. But even as an indicator we are still in the guessing game; no inductive link is allowed. We simply guess, or conjecture, that the better corroborated theory is nearer the truth (1974, p. 1011). 'But I can examine my guess critically, and if it withstands severe criticism, then, this fact may be taken as a good critical reason in favour of it' (1972, p. 234). This is paramount to admitting that there is no formal link between corroboration and verisimilitude, and this is the conclusion that Popper reluctantly has to admit. (In the 'Supplementary Remarks' to *Objective Knowledge* he apologises for 'having made some very serious mistakes in connection with the definition of verisimilitude' (1979, p. 372).) But although the attempt to define verisimilitude has been shown to be unsatisfactory Popper maintains that it does not follow that it should be abandoned. We do not abandon science because our theory of science is wrong; and science without verisimilitude is irrational. Here is an interesting example from the great advocate of falsification himself of reluctance to reject a theory that has not withstood severe criticism.

Popper takes up this issue again in the 1983 introduction to *Realism and the Aim of Science*. Although he readily admits his failure to formulate a satisfactory definition of verisimilitude, he is adamant that the idea of 'approach to the truth' is intuitively correct, and that it is difficult to give a rational account of scientific knowledge without it. He supports his contention with examples from the development of the theory of the solar system, and the development of ideas about heredity from Darwin to Mendel (1983, p. xxxvi). Thus he continues to maintain that, although

> we *cannot* ever have . . . good arguments in the empirical sciences for claiming that we have actually reached the truth . . . we *can* have good arguments for preferring one competing theory to another one *with respect to our aim of finding the truth, or of getting nearer to the truth*. Thus I hold that we *can* have good arguments for *preferring* – if only for the time being – T_2 to T_1 *with respect to verisimilitude*. (1979, p. 372)

He emphasises that the *criterion of progress* referred to earlier in terms of content is not a claim that one theory is in fact nearer the truth than another (e.g. Einstein's compared with Newton's) but rather it is 'an appraisal of the *state of discussion* of these theories' (1979, p. 372).

Critics of Popper find this hard to swallow, for, they maintain, if truth or verisimilitude is the aim of science, one is thwarted from the very outset from arriving at any rational claim to the truth. Truth, or approach to the truth, which is the goal of the whole enterprise of conjecture and refutation, has, it seems, no determinate connection to the very efforts one undertakes in order to secure it. Jonathan Lieberson attempts to sum up Popper's philosophy of science in the following advice given to a scientist:

> You have created a hypothesis designed to solve a significant scientific problem. You must now test your idea by trying to falsify it; induction is non-existent, and this is the sole 'rational' way of trying to get closer to the truth. But you must also be aware that your testing, however prolonged, will never provide you with a scintilla of rational evidence for thinking you are advancing toward, or receding from, the truth, the goal of your enquiry; it cannot even do so for the claim that you have found out your hypothesis is empirically false. Nevertheless you must continue testing. And when you stop testing, remember that your tests will have given you no help in rationally deciding whether your hypothesis is true or not; you must simply guess that it is true, or discard it, because it contradicts some other hypotheses which you have decided are true, once more without any rational grounds of any kind. (Lieberson, 1983, p. 44)

Induction Again?

Those of Popper's critics who agree that scientific theories can be rationally compared maintain that an inductive link must be maintained if we are to uphold the rationality of science. In maintaining the intuitive idea that Einstein's theory has greater verisimilitude than its predecessors (such as Newton's) that make less successful predictions, Popper comes close to making that very admission when he says that

> there may be a 'whiff' of inductivism here. It enters with the vague realist assumption that reality, though unknown, is in some respects similar to what science tells us or, in other words, with the assumption that science *can progress towards greater verisimilitude* (1974, p. 1193, note 165b)

This appears to be an admission on Popper's part that verisimilitude only really makes sense in inductive terms. O'Hear, commenting on this passage, suggests that it is a concession to the inductivists and that it entitles us 'to rationally rely on the findings of science' (O'Hear, 1980, p. 67). Popper's reply to this would be to distinguish between reliance and reliability. We may, and indeed do, rely on scientific theories, but that is not to say that they are reliable, and we should therefore always try to 'foresee the possibility that something may go wrong' (1974, p. 1026).

Newton-Smith, in his book *The Rationality of Science* (1981), maintains that far from being a 'whiff' of inductivism, it is a full-blown storm, and therefore 'if we concede a role to induction here there is no reason not to admit inductive argument right from the start. If we do this we lose what was unique and interesting in Popper: namely, the jettisoning of induction' (p. 70).[14] Thus Newton-Smith depicts Popper as the *Irrational Rationalist* because he at one and the same time dismisses all inductive inference as irrational but eventually has to resort to it to justify his contention that science is a rational activity the goal of which is, if not truth, increasing verisimilitude. Jonathan Lieberson arrives at the same conclusion when he writes:

> Popper's philosophy of science contains many suggestive and stimulating ideas. But for all its breadth and sweep and occasional flashes of insight, it still seems to me fundamentally disfigured, in large part by its initial rejection of induction. It is, in words that were used by Santayana in another context, 'capable of assimilating a great deal of wisdom, while its first foundation is folly'. (1983, p. 44)

Putnam makes the same point rather more graphically when he says:

> When a scientist accepts a law, he is recommending to other men that they rely on it . . . Only by wrenching science altogether out of the context in which it really arises . . . can Popper even put forward his peculiar view on induction. Ideas are not *just* ideas; they are guides to action. (Putnam, 1974, p. 222)

O'Hear, commenting on this passage, concludes his comprehensive and detailed assessment of Popper by suggesting that Wittgenstein would never have made the mistake of wrenching ideas out of their social and practical context. This is a somewhat ironic comparison because for Wittgenstein there could *be* no problematic philosophical theory of science (philosophy dissolves problems), whereas Popper recognises problems as fundamental to the human evolutionary scene, and he tries to provide a rational account of the way we move progressively through problem situations. Of course, Popper can be criticised from a Wittgensteinian standpoint, that the task of philosophy is descriptive rather than problem solving, but we will pursue this line in relation to the philosophy of science in our next chapter.

Referring to corroboration in relation to verisimilitude Putnam suggests that Popper's talk about the 'best theory' implies an inductivist quaver. The suggestion here appears to be that inductivism is deeply written into our conceptual scheme and we can hardly think or act, let alone do science, without it. Popper might well reply that this is indeed the case, and for that reason a 'gestalt switch' seems needed — a complete break with the traditional way of looking at these things and with the common-sense theory of knowledge (1974, p. 1044). Such a switch in our conceptual scheme by which we purge our minds of induction not only has fruitful consequences in terms of a theory of rationality, but also for practical action; in terms of the progress of scientific knowledge, and in social theory — looking for the unintended and unexpected consequences of our decisions.

Our discussion of the elusive concept of truth has highlighted the main thrust of criticism levelled against Popper, and it brings us back to the fundamental question of induction. Can we speak about truth without some justificationary inductive principle? Is the step from corroboration to verisimilitude an inductive one or not? Are we confronted here with a fundamental point of principle or are we engaged in a mere verbal quibble? Popper would be the first to agree that arguments about terminology are unimportant, and maybe he has to concede that what he conceives of as a *guess* or a *rationally entertained speculation* is as well described as an inductive link.

Perhaps we may summarise Popper's position as follows. Popper agrees with Hume that induction is irrational. However, science is rational, and therefore it cannot proceed inductively. Although the aim of science is increasing verisimilitude, and explanatory power, all knowledge is fallible and therefore all 'truths' are held provisionally. Thus the method of conjecture and attempted refutation is the only rational way of proceeding. Where a 'whiff' of induction appears, we must treat it as a 'guess' or a 'rationally entertained speculation'.

Lakatos has made an instructive distinction in this respect between three independent aspects of induction to which Popper has addressed himself (1974, pp. 258-61).

(1)   *The inductive logic of discovery* according to which theories result from accumulated observations. Due not least to Popper, this is 'now only taken seriously by the most provincial and illiterate' (such as the compilers of the Bangor *Prospectus*!).

(2)   *A priori probabilistic inductive logic* or confirmation theory, according to which certainty in science is replaced by a degree of probability (according to the probability calculus). Popper has achieved 'complete victory' in this area too, despite the attempts of the proponents of inductive logic.

(3)   *A synthetic inductive principle* connecting theory appraisals (content and corroboration) with verisimilitude. This third aspect is, as we have indicated, the weak point in the Popperian onslaught against induction, and here even Popper himself has at one point admitted to a 'whiff' of induction, or an 'inductivist quaver'. Lakatos points out that what is at stake here is the subtle difference between constructive fallibilism and profound scepticism, 'with all its evil consequences, like relativism, irrationalism, mysticism', and, we might add, Feyerabend's epistemological anarchism.

> By refusing to accept this 'thin' metaphysical principle of induction Popper fails to separate rationalism from irrationalism, weak light from total darkness. Without this principle Popper's 'corroborations' or 'refutations' . . . would remain mere honorific titles awarded in a pure game. With a *positive* solution to the problem of induction, however thin, methodological theories of demarcation can be turned from arbitrary conventions into rational metaphysics. (Lakatos, 1974, p. 261)

There is no doubt that Popper's fallibilism is informed by a positive and optimistic view of human knowledge, and it is this optimism that makes his philosophy so attractive compared with the gloomy scepticism characteristic of much Anglo-Saxon philosophy since Hume. The admission of an inductive link between corroboration and verisimilitude does not imply justificationism in any final or absolute sense, but it does allow for the necessary provisional justifications which we need if we are to think and act.

It is at this point that the tension in Popper's philosophy, between logic and method, between form and practice, is most evident. Putnam anticipated Kuhn in maintaining that 'practice is primary' (Putnam, 1974, pp. 237-9), but for Popper the *logic of the situation* must inform the practice.

Before we turn to the influential analysis of science offered by Kuhn as a corrective and alternative to the Popperian account, let us examine in greater detail some of the wider implications of Popper's philosophy of science, and in particular its bearing on religion.

The Implications for Religion

The Spirit of Enquiry

We have outlined the powerful and long-lasting influence that the positivistic theory of science had on religion, and on the unquestioning submission of religious apologists to this Received View. We reject this theory of science, not because it threatens religion or because it does not support our prejudices, but because it is a fallacious doctrine which, by demolishing all metaphysics, demolishes the very science it set out to account for. Both science and religion are human enter-prises, deriving from a common source and sharing the same problems associated with questions arising from man's puzzlement about his world. The metaphysical basis is common to both.

The Popperian view of science, on the other hand, has remarkably fruitful consequences for religion, and yet philosophers of religion have never exploited them. The main reason for this is that the Received View of science has had such a powerful hold that Popperian ideas were read as part of the positivist programme. It is not surprising that religious thinkers should hold this view if philosophers of science themselves continued for a long time to see Popper as an advocate of a brand of empiricism that was essentially positivistic. A further reason for the neglect of Popper by philosophers of religion lies in the bewitching influence of Wittgenstein, whose philosophy of language appeared to offer a safe haven for the religious talk that had been forbidden by the followers of the earlier Wittgenstein. This is doubtless the reason why philosophers of religion have readily latched on to the Kuhnian view of science, as we shall indicate in the next chapter.

The overall point to be made about Popperian philosophy of science in relation to religion is its extraordinarily wide, almost cosmic, sweep. Science is set in an evolutionary context, physical, biological and cultural, and even extends into a third world of objective ideas which transcends the worlds of the physical and the mental. This is perhaps the point to exploit first as we consider the implications of Popperian philosophy for religion. If it is important for a theory of science to be informed by a descriptive epistemology arising from an evolutionary view, how much more important that religion should be considered against the broad evolutionary canvas. It is largely through the work of Popper that the philosophy of science has broken out of a narrow and uninformed parochialism concerned in the main with the internal structure of theories. Similarly, philosophers of religion have been over-occupied with the examination of 'religious language', and the structure of theological statements concerning the existence of God.

Religion, it is said, is about the relationship between man and God, and, we would add, the world. Consequently, if we are to grow in our understanding in these matters, then we must set our debate in the context of an informed view of the whole evolutionary story. Popper's unflagging defence of realism and rationality arises from the fundamental assumption of science that the universe, although complex and mysterious, is in some degree intelligible. In his *Science and the Modern World* (1938), Whitehead suggested that the view of the world as intelligible arose out of the Judaeo-Christian doctrine of creation. A universe created by an intelligent Being can be the object of investigation by intelligent beings. The stagnation of science for close on 2,000 years was in no small measure due to the dominant influence of Aristotelian philosophy which depicted the universe as uncreated and eternal and therefore not open to investigation in terms of material causes. Thus the theist, who sees the world as God's creation, shares Popper's fundamental faith in man's ability, albeit painfully and slowly and erringly, to unravel some of the secrets of the created order. In depicting evolution as a problem solving affair, Popper is supporting the thesis that it is part of man's nature to try to understand his world, and his rational reconstruction of this attempt is an essential part of the theist's endeavour to examine the relationship between man, God and the world.

Arising out of the scientist's exploration of the world with his theories is a sense of awe and respect and wonder which is akin to the religious sense, and which moves Popper to describe science as 'one of the greatest spiritual adventures that man has yet known' (Popper, 1957, p. 56). When Popper writes in *The Logic of Scientific Discovery*, 'It is not truisms which science unveils. Rather, it is part of the greatness and the beauty of science that we can learn, through our own critical investigations, that the world is utterly different from what we ever imagined' (1968a, p. 431), he is in accord with the religious view which holds that the world is sustained by a reality of a different and deeper order from that which our everyday, common-sense experience would suggest. Whether it is primitive man puzzled by the bent stick under the surface of the water, or Newton enchanted by the constituent colours of the visible light spectrum, or Rutherford's incredulity that alpha particles were deflected by a sheet of gold foil, or Eddington's excitement at the deflection of light by gravity, or Stephen Hawking's conjectures about 'black holes', science reveals hidden realities behind the world of appearances. Likewise the lily of the field, which is beautiful to behold, when examined under the microscope of science, reveals a structure of breathtaking beauty and complexity. It is at this level that science and religion share common ground, and the panoramic sweep of Popper's philosophy of science, and the astonishing range of his enquiries, help to enlarge our vision and expand our minds as we contemplate the fundamental questions concerning the relationship between man, God and the world.

Although Popper seldom overtly refers to religion, and he is not, as Magee comments, 'what people usually mean by a religious man', (1973, p. 37), his attitude to religion is certainly not dismissive or hostile. In a conversation with Eccles, his collaborator in *The Self and its Brain*, Popper admits an openness and sympathy for Eccles' religious attitude regarding supernatural creation, the soul and survival beyond bodily death, and he maintains that 'in spite of disagreeing we take seriously and respect each other's views' in these matters (Popper and Eccles, 1977, Dialogue XI). And when Popper contemplates the incredibly improbable emergence of self-consciousness from inanimate matter he admits with Eccles that we are confronted by a mystery for which evolutionary theory offers no explanation. This is a far cry from the nineteenth-century confidence in scientific explanation, which it was thought would dissolve all mysteries, and echoed by Laplace's reply to Napoleon regarding religious explanation: 'we have no need of this hypothesis'.

Given that science and religion share common ground at the level of man's puzzlement and of his attempt to make sense of his world, let us examine in more detail some of the more specific points in Popper's philosophy which touch more directly on religious matters.

Metaphysics and Demarcation

We have observed how the positivists extended their inductive criterion of demarcation between science and non-science into a general criterion of meaning, thus legislating between sense and nonsense. We cannot over-emphasise the profound and widespread influence of this view, which, even though it was short-lived among philosophers, still forms the basis of contemporary man's attitude to religion. H. D. Lewis, in his *Philosophy of Religion* (1965), suggests that Ayer's characterisation of religion as meaningless nonsense reflects 'the attitude of many who are indifferent or opposed to religion today'. He continues, 'It is not so much that they find themselves unconvinced but that most that religious people say and do is altogether without meaning to them' (p. 75). Doubtless there is a complex of social and economic factors involved here, but at root the attitude is based on a philosophical presupposition — that science is about *sense* and religion is about *non-sense* and therefore nonsense.

Popper never made the mistake of trying to pontificate about meaning, between what could be said, and what could not be said, meaningfully; and he did more than any other philosopher of his time to rehabilitate metaphysics as indispensable to man's attempt to understand the world and advance his knowledge.

Popper's original deference to metaphysical speculation was the result of his recognition of the indispensability of metaphysics for science, not only as ineradicable from empirical theories, but more importantly as the characteristic of the forerunners of testable scientific theories. He refers to these speculative, but untestable, ideas as *metaphysical research programmes.*[15] Positivism always opposed such speculations. These research programmes play a crucially important role in the development of science; if and when they become testable they take on the character of scientific theories. We have already noted Popper's attitude to Darwinism as a metaphysical research programme. In the 'Metaphysical Epilogue' to the *Postscript* he lists, as examples, ten of the more important metaphysical research programmes that have influenced the development of physics since the days of Pythagoras and Heraclitus; from *Parmenides' Block Universe* to the *Statistical Interpretation of Quantum Theory* (1982a, pp. 162-4). He maintains that

> Such research programmes are, generally speaking, indispensable for science, although their character is that of metaphysical or speculative physics rather than of scientific physics. Originally they were all metaphysical, in nearly every sense of the word (although some of them became scientific in time); they were vast generalisations, based upon various intuitive ideas, most of which now strike us as mistaken. They were unifying pictures of the world — the real world. They were highly speculative; and they were, originally, nontestable. Indeed they may all be said to have been more of the nature of myths, or of dreams, than of science. But they helped to give science its problems, its purposes, and its inspiration. (1982a, p. 165)

The important feature of these metaphysical speculations is that they

> proved *susceptible to criticism* — that they could be critically discussed. It was a discussion inspired by the wish to understand the world, and by the hope, the conviction, that the human mind can at least make an attempt to understand it. (1982a, p. 172)

The schism in physics today results from a clash between the classic

Faraday-Einstein-Schrödinger programme and that resulting from quantum mechanics. Those who maintain that there is no clash, that the Einsteinian conception has been ousted, are left with an instrumentalist quantum theory that is no theory because it has little explanatory power. Popper's argument in the *Postscript* is an attempt to resolve the conflict by relating three metaphysical theses: indeterminism, realism and objectivism in terms of what Popper calls 'propensity'. The details of this interesting and ingenious account need not detain us here, but its significance lies in the admission that explanatory physical theory, in the last analysis, rests on deep, untestable, metaphysical principles. Popper describes his programme as more in the nature of a dream, which

> tries to give a coherent view of the physical world — a physical world which is no longer a strait-jacket for its physical inhabitants, not in a cage in which we are caught, but a habitat which we may make more habitable, for ourselves and for others (and which incidentally, we are about to make uninhabitable for our children by what we proudly call 'the peaceful use of atomic energy').[16] (1982a, p. 199)

At this point Popper admits to a change in his original attitude concerning demarcation, which was introduced to distinguish between science and pseudo-science, and according to which only refutable scientific theories could be claimants to truth. He writes:

> I no longer think, as I once did, that there is a difference between science and metaphysics regarding this most important point. I look upon a metaphysical theory as similar to a scientific one . . . *as long as a metaphysical theory can be rationally criticized*, I should be inclined to take seriously its implicit claim to be considered tentatively as true. (1982a, p. 199)

Popper is not here identifying empirical science with metaphysics, but suggesting that although metaphysical theories are irrefutable they can be critically evaluated. The question arises, in what does a rational appraisal or evaluation consist? Popper's answer is:

> If a metaphysical theory is a more or less isolated assertion, no more than the product of an intuition or an insight flung at us with an implied 'take it or leave it', then it may well be impossible to discuss it rationally. But the same would be true of a 'scientific'

theory. Should anybody present us with the equations of classical mechanics without first explaining to us *what the problems are* which they are meant to solve, then we should not be able to discuss them rationally – no more than *The Book of Revelation*[17] . . . in other words, any rational theory, no matter whether scientific or metaphysical, is rational only because it ties up with something else – because it is an attempt to solve certain problems: and it can be rationally discussed only *in relation to the problem situation* with which it is tied up. (1982a, p. 200; see also 1972, pp. 198-9)

In his discussion of the problem of the 'Irrefutability of Metaphysical Theories' in *Conjectures and Refutations* (Ch. 8), Popper quickly disposes of the false notion that the truth of a theory may be inferred from its irrefutability; two irrefutable but incompatible theories cannot both entail truth. But falsity may be inferred by rational argument, given the conditions he specifies concerning the problem situation and so on. He lists five philosophical (metaphysical) theses: determinism, idealism, irrationalism, voluntarism and nihilism, each of which, although irrefutable, he considers false. His conclusions regarding these matters might be disputed (there are those, for instance, who still cling to the metaphysical thesis of determinism), but that he does bring a whole weight of rational argument to support his contentions cannot be disputed. Popper suggests that specific features of a rationally arguable metaphysical programme are simplicity, coherence, unifying power, intuitive appeal and, above all, fruitfulness. Are these not the marks of the theories of the 'heroic' scientists we listed earlier?

It is of crucial importance to see the direction in which Popper's thought has moved, from his original criterion of demarcation between science and non-science (in order to establish the characteristics of the growth of scientific knowledge), to a demarcation *within* metaphysics between criticisable and non-criticisable theories (see 1968b, p. 95).

The implications of this for philosophy in general, and for religion in particular, are profound. In the first place, Popper's original principle of demarcation does not automatically exclude the propositions of religion from significant discourse (as did the positivist criterion of meaning). The fact that statements about God, for instance, cannot be falsified, does not render them any more nonsensical than the metaphysical forerunners of scientific theories. (That is not to say that empirical evidence has no bearing on the *decisions* of religious people concerning their beliefs. Although two men confronted by the same facts, of human suffering say, can draw opposite conclusions concerning the existence of a loving God, they can continue to discuss rationally their problem in the light of the evidence.) Here was a theory of science that at least allowed religious talk to get going, whereas positivism ruled it out of court from the start.

But the problem for the sort of metaphysical speculation that religious assertions involve that has preoccupied philosophers of religion concerns their truth. Religious apologists are still bedevilled by questions such as 'How can we know whether religious statements are true or not?' or 'What is the nature of religious truth?' And because answers to these questions have not been forthcoming they have taken refuge in the reductionism we discussed in the last chapter, and in the sort of relativism and fideism we shall examine in the next chapter. Popper's fundamental contention is that we cannot infer the truth of *any* theory, whether scientific or metaphysical, but we can have some idea as to its falsity. And this is the key to the rational appraisal or evaluation of *any* theory, whether scientific or metaphysical. Thus the philosophers of religion who endeavour to establish the *truth* of religious assertions are as misguided as philosophers of science who attempt to construct a theory of verification for scientific theories. Rational discussion of a theory is an attempt to evaluate it critically in the context of a particular problem situation.

It does not follow from Popper's admission of rationality within metaphysics that nonsense cannot be talked. Anyone, scientist, philosopher or theologian, can string words into sentences and 'fling' them at us with an implied 'take it or leave it' and thus prohibit rational discussion. We generally recognise an uncriticisable or irrational assertion when we meet it.

It is our contention that theological speculation can be, and generally is, a rational affair. (That is not to say, of course, that theological rubbish is never produced — and published!) Theological statements, if satisfactorily formulated and related to a particular problem situation, say about the relationship between man, God and the world, can be rationally and critically discussed, with particular regard for their fruitfulness. Although truth cannot be established, falsity (or absurdity) often can.[18]

Take, for instance, John Hick, who in some of his publications discusses the dilemma modern man is in, confronted as he is by a number of conflicting world religions, each a claimant to truth.[19] In the face of this he argues for what he calls a God-centred 'Copernican revolution' according to which each religion is accepted as holding a phenomenal part of the noumenal divine behind all of them (Hick, 1980, pp. 5-6 and 51-2). The basic assumption behind this idea is that because there are conflicting claims to religious truth each religion ought to be construed as an equally valid apprehension of the one divine reality until such time as 'eschatological verification' will resolve the issues. This is Hick's primary motive for his insistence on the mythological nature of the Christian doctrine of the incarnation: the doctrine cannot be 'literally true' because it would preclude the soteriological effectiveness of other religions.

But we can no more treat religion as a whole than we can treat science as a whole. Granted that in religion, as in science, we cannot isolate individual beliefs or theories from the whole body of which they are a part, we can only subject particular truth claims to critical examination. There is no other way of proceeding. On Hick's model it may turn out in the final eschatological analysis that only one specific theology was valid. We have already noted the irrelevance of Hick's notion of 'eschatological verification'. The only rational procedure is to subject individual theories to critical examination and attempt to weed out falsity.

Thus it does not follow that because there are several brands of religion, a number of credal systems, we should therefore look upon them as of equal value, and not attempt to assess one at the expense of another. To suggest, as Hick does, in the face of religious pluralism, that it is impossible to assess rationally conflicting claims to religious truth, is both absurd and dangerous; it allows any belief or action if it is performed 'in the name of God'. Religious tolerance is one thing and is much to be commended in the face of bigotry and sectarianism, but it does not follow that religious claims to truth are incommensurable. The Christian and the Muslim and the Hindu, for example, make objective, and therefore rationally arguable, claims. Agreement is by no means easy and the cultural factors weigh heavily, but if the metaphysical theories which form the basis of natural science can be rationally assessed, then by the same token so can the irrefutable theories of religion. The evaluation is always critical, aiming at the elimination of error rather than the establishment of absolute or final truth. And as with all rationally arguable metaphysical programmes, we have regard for simplicity, coherence, unifying power, consistency and, above all, for fruitfulness, which is of overriding importance. Thus a religion which encourages child sacrifice, or which perpetuates a caste system or the practice of suttee, or which enslaves its adherents, or which condones violence as a means to an end, can be rationally criticised in terms of its consequences, or as Popper would put it, in terms of its fruitfulness. This is no resort to the pragmatic theory of truth, which equates truth with usefulness, because the aim is not the establishment of truth, but the elimination of error. The evaluation is often complex and difficult and sometimes inconclusive as it is in science.

Popper concludes his 'Metaphysical Epilogue', which he likens to a picture or a dream rather than a testable theory with the words

> perhaps . . . we may find a criterion of demarcation *within metaphysics*, between rationally worthless metaphysical systems, and metaphysical systems that are worth discussing, and worth thinking about. The proper aspiration of a metaphysician, I am inclined to say, is to gather all the true aspects of the world (and not merely its scientific aspects) into a unifying picture which may enlighten him and others, and which one day may become part of a still more comprehensive picture, a better picture, a truer picture. The criterion, then, will be fundamentally the same as in the sciences. Whether a picture is worth considering depends, I suggest, upon its capacity to provoke rational criticism, and to inspire attempts to supersede it by something better (rather than upon its capacity to create a fashion, to be supplanted presently by a new fashion, or upon claims to originality or finality). (1982a, p. 211)

The theist will want to insist that if we are to 'gather up all the true aspects of the world' then we must include rational attempts to explain the world in relation to its Creator. But if the attempt is to remain within the (Popperian) concept of rationality, then it must undergo criticism and be open to change and growth, rather than claim finality and remain a once-for-all static dogma. To the question of truth in religion we must now turn.

The Retreat from Authority

The abandonment of the claim to have discovered truth in science has profound and far-reaching implications for religion. For the religious apologist has long laboured under the illusion that the fundamental difference between the claims of science and the claims of religion concerns the question of truth. Whatever the strength of his conviction, the religious believer cannot 'prove' his assertions concerning the relationship between man and God, whereas the scientist can adduce evidence which, if it does not prove his theories, renders them highly probable.

H. D. Lewis admirably summarises this mistaken view of science versus religion when he writes:

> We must be content with probability, although in many matters this is of a very high order indeed *and gives us all the certainty we need.*[20] Religious people often claim certainty, but the truth is that we cannot have even a degree of probability in religious matters. (1965, pp. 73-4)

Although, as we have observed, the Received View of science admitted a degree of fallibilism by substituting high probability for certainty, it is Popper who has convincingly demonstrated that the connection between high probability and truth is an illusion. All our knowledge is fallible. Whatever the strength of our evidence, however successful a theory has been up to now, we could be wrong. Fallibilism is an essential feature of World 3, which contains true and false theories. Popper is fond of quoting the pre-Socratic philosopher Zenophanes who recognised that all our knowledge is guesswork.

> The gods did not reveal from the beginning,
> All things to us, but in the course of time
> Through seeking we may learn and know things better.
>
> But as for certain truth, no man has known it,
> Nor will he know it; neither of the gods,
> Nor yet of all the things of which I speak.
> And even if by chance he were to utter
> The final truth, he would himself not know it;
> For all is but a woven web of guesses.

(1972, p. 26)

It was within the very heart of modern science that Popper discerned the fallibility of all our knowledge. Newtonian physics was the most remarkable and successful scientific theory ever conceived. It influenced a whole era of civilisation and provided the theoretical basis for the discoveries of the industrial revolution. And a new theology grew out of it: the clockwork universe of deism. And yet the theory was wrong. Why? Because it came not from the gods but from man, from the mind of Newton. Knowledge is man-made, not impressed on us by the world, or revealed by the gods.

The recognition that all our judgements are human and therefore fallible is not a denial of the objective nature of absolute truth. Truth remains the goal of human enquiry; it is the indispensable regulative principle without which we must lapse into relativism and irrationalism. And although we can never positively identify it, we do have a pretty good idea at times when we are getting a little nearer the truth.

In 'On the Sources of Knowledge and Ignorance' (1972, Introduction), Popper emphasises the profound practical consequences for our lives which result from the theoretical basis we adopt regarding the nature of knowledge and truth. He enlists the support of Russell who perceived the relationship between epistemology and the ordering of the institutions that form a society. The central question here concerns the nature of authority, for this is the link between the Cartesian rationalists and the Baconian empiricists. Although the epistemologies of Bacon and Descartes were conceived as a revolt against the traditional authorities of Aristotle and the Bible, they too were lured by the appeal to authority: the one to the authority of the senses and the other to the authority of the intellect. But each of these epistemologies was based on a myth: the myth that *the truth is manifest*. According to this 'optimistic epistemology', once the truth is revealed to us we immediately recognise it for what it is (1972, p. 5). If truth is manifest, either as a result of the *veracitas dei* of Descartes or the *veracitas naturae* of Bacon, then what is the source of our ignorance? The answer lies in man's sinfulness, or evilness, or wilfulness, or prejudice, and what is required is a purging of the mind of man from such pernicious influences in order that he may perceive the naked truth.

The optimism generated by this false epistemology has had remarkably fruitful consequences. It was 'the major inspiration of an intellectual and moral revolution without parallel in history'. Popper continues:

> It encouraged men to think for themselves. It gave them hope that through knowledge they might free themselves and others from servitude and misery. It made modern science possible. It became the basis of the fight against censorship and the suppression of free thought. It became the basis of the nonconformist conscience, of individualism, and of a new sense of man's dignity; of a demand for universal education, and of a new dream of a free society. It made men feel responsible for themselves and for others, and eager to improve not only their own condition but also that of their fellow men. It is a case of a bad idea inspiring many good ones. (1972, p. 8)

But the theory – that the truth is manifest – has also had disastrous consequences for it is the basis of almost every kind of fanaticism and authoritarianism; for those who do not see the manifest truth must be coerced by the authority of those who do.

The whole tone of Popper's thought is consistently anti-authoritarian. We cannot justify our knowledge by appeal to some ultimate authoritative source of true knowledge, and no man's authority can establish truth by decree. Truth is above human authority. Truth, far from being manifest, *is hard to come by* (1972, p. 6).

Our hope lies not in unveiling revealed truths which impress themselves upon us by their authority, but in the detection and elimination of error through the criticism of our theories, conjectures and guesses. We must not embark on the fruitless search for ultimate sources of true knowledge. Rather, we attempt to approach the truth through the elimination of error and falsehood, and 'in searching for the truth it may be our best plan to start by criticizing our most cherished beliefs' (1972, p. 6).

As Popper observes, Kant boldly carried this idea into religion when he wrote, 'in whatever way the Deity should be made known to you, and even if He should reveal Himself to you: it is you . . . who must judge whether you are permitted to believe in Him, and to worship Him' (1972, pp. 26 and 182). So in religion, as in science, human, critical judgement is the arbiter. And if we apply the Popperian method of critical rationalism there can be no place for fundamentalism; no uncritical, unquestioning acceptance of a dogmatic tradition invested in the Word, the Institution, the Leader. Christian fundamentalists who elevate the Bible, or the Pope, or the closed individual conscience, beyond all criticism lay themselves open to all the consequences of irrationalism. This is not to deny the importance of tradition, for as Popper constantly reminds us, most of the sources of our knowledge are traditional. Without tradition knowledge would be impossible. But 'every bit of our traditional knowledge (and even our inborn knowledge) is open to critical examination and may be overthrown' (1972, p. 28).

Tradition in religion (as in science) is of supreme importance. We do not disregard it, nor do we easily reject it. But we must not identify even the most hallowed of well-criticised tradition with the ultimate truth. Thus religious doctrines, credal statements, must never be treated as absolute, eternal verities. Even the statement 'God is love' which appears to be an unshakeable Christian conviction must not be glibly accepted, because the concept of love is not a static but a dynamic affair and open to critical reassessment.[21]

If we reject the notion of religion furnishing its adherents with once-for-all static and eternal verities then there is room for growth in theology. But if our knowledge of God and his relationship with man and his world is to grow then our theology must be open and not closed. It must be open not just to the adjustments of linguistic flux, but to fresh insights, fresh conjectures to be severely criticised.

This opens up the whole question of conviction and faith. The religious man is frequently depicted, generally by those outside his faith, as a man without doubt, a man convinced by his beliefs. But the degree of an individual's conviction is a matter of psychology and it has no bearing on the truth or falsity of the object of his belief. Only when the inner conviction becomes a publicly criticisable credo can it lay claim to rationality. What we must not do is to confuse religious faith with conviction. Austin Farrer has pointed to the distinction between faith and belief. Faith involves a decision to act, to behave in a certain way. To test one's belief about the soundness of the plank by attempting to cross it. Thus the religious man can have a strong faith, a firm commitment, but remain uncertain, unsure, and live with a varying degree of scepticism and doubt. Austin Farrer comments:

> Suspense of judgement that results in refusal to act, is practical atheism . . . not only is it true that all agnostics are either believers or disbelievers; equally it is true that all believers and disbelievers are agnostics: for the man who professes to have absolute knowledge of God in this world is a strange beast. The question then is . . . Am I prepared to bet on my opinion by making my life answer it? (Farrer, 1972, p. 8)

Bishop Stephen Neill, owning to a radically sceptical temper, admits that 'I still wake up on about three mornings a week, saying, "Of course it couldn't possibly be true" ' (Neill, 1977, p 88). But he remains a committed, confessing Christian without doing damage to his integrity.

The man of faith, if he is rational, will remain critical and open to change. There is nothing particularly revolutionary in this idea with regard to religion. After all, the central challenge presented by the early followers of Jesus concerns *metanoia* or change (a word unfortunately translated as *repentance*) (Acts 2.38, AV). The challenge was as much to a previously uncriticised tradition as it was to the individual conscience.

John Eccles has testified to the liberating effect of Popper's teachings for the working scientist (Eccles, 1974, p. 350). Likewise the admission of fallibilism in the place of authority can have a liberating and stimulating effect on the religious man. Far from resenting criticism one should see it as an invaluable aid, not only to the growth of knowledge, but to spiritual growth. Magee comments:

> The man who welcomes and acts on criticism will prize it almost above friendship: the man who fights it out of concern to maintain his position is clinging on to non-growth. Anything like a widespread changeover in our society towards Popperian attitudes to criticism would constitute a revolution in social and interpersonal relationships − not to mention organizational practice . . . (Magee, 1973, pp. 39-40)

Truth remains the goal for both scientist and theologian. The way towards that goal is common to both. It is the way of criticism and the admission of fallibilism. Julian Huxley pointed out over fifty years ago that

> a radical difference of outlook obtains between science and religion. An alteration in scientific outlook − for instance, the supersession of pure Newtonian Mechanics by relativity − is generally looked on as a victory for science, but an alteration in religious outlook − for instance, the abandonment of belief in the literal truth of the account of creation in Genesis − is usually looked on in some way as a defeat for religion. Yet either both are defeats or both victories − not for partial activities, such as religion or science, but for the spirit of man. (Huxley, 1931, p. 21)

There is a sense in which criticism (or rationality) is to be prized above all Christian virtues, even above the Pauline trio of faith, hope and charity, for how are we to decide on the application of these virtues without a rational assessment of the instance in question? Take the greatest of all Christian virtues, love. Popper holds 'that he who teaches that not reason but love should rule opens the way for those who rule by hate' and he demonstrates with a harmless example how even this most constructive human emotion is usually unable to decide a conflict.

> Tom likes the theatre and Dick likes dancing. Tom lovingly insists on going to a dance while Dick wants for Tom's sake to go to the theatre. This conflict cannot be settled by love; rather, the greater the love, the stronger will be the conflict. There are only two

solutions; one is the use of emotion, and ultimately of violence, and the other is the use of reason, of impartiality, of reasonable compromise. All this is not intended to indicate that I do not appreciate the difference between love and hate, or that I think that life would be worth living without love. (And I am quite prepared to admit that the Christian idea of love is not meant in a purely emotional way.) (1966, Vol. 2, p. 236)

But love, as the mainspring of human action, unregulated by the critical faculty of reason, leads as often as not to an appeal to violence and brute force as the ultimate arbitrator in human affairs. Popper's conclusion 'that no emotion, not even love, can replace the rule of institutions controlled by reason', will be pursued in the next section.

The question of course arises, 'Is it a rational decision to behave rationally?' Popper admits that the decision to adopt a rational attitude cannot be justified by reason. A non-justificationary philosophy cannot resort to the justification of its own principles. The question of whether or not rationalism is a matter of blind faith, an irrational commitment, has been the subject of considerable debate. Bartley, in *The Retreat to Commitment* (1962) and 'Rationality versus the Theory of Rationality' (1964), attempts to extricate Popper from what he calls the *tu quoque* of the irrationalist. But O'Hear (1980, pp. 147-53) maintains that the rationalist can simply reject the irrationalist's *tu quoque* on the grounds that a justification of rationality is *logically* unobtainable. He argues that a commitment to rationalism need not be based on an irrational faith because there can be no standard against which rationality as a general principle can be assessed. It is senseless to make a demand for what is logically unobtainable. There is something spurious about the attempt to produce a reasoned defence of reason. It is impossible to criticise a rational theory of criticism. What Popper has attempted to do is to identify the characteristic of the growth of scientific knowledge and to suggest that we abandon this method of criticism and the admission of fallibilism at our own peril. It is more in the nature of a moral decision which is tested against the consequences of irrationalism. What we are suggesting is that the religious man would do well to adopt the same attitude of critical rationalism and be prepared with Popper

to give up the idea of ultimate sources of knowledge, and admit that all knowledge is human; that it is mixed with our errors, our prejudices, our dreams, and our hopes; that all we can do is to grope for truth even though it be beyond our reach. We may admit that our groping is often inspired, but we must be on our guard against the belief, however deeply felt, that our inspiration carries any authority, divine or otherwise. If we thus admit that there is no authority beyond the reach of criticism to be found within the whole province of our knowledge, however far it may have penetrated into the unknown, then we can retain, without danger, the idea that truth is beyond human authority. And we must retain it. For without this idea there can be no objective standards of inquiry; no criticism of our conjectures; no groping for the unknown; no quest for knowledge. (1972, p. 30)

Open and Closed Societies

This brings us to a consideration of Popper's social and political philosophy, which is consistent with the epistemology from which his philosophy of science is derived, and which bears directly on the social and political dimension in religion. The remarkable thing about Popper's social and political ideas lies in the way in which they are derived from his epistemology in general and his philosophy of science in particular. The fundamental importance of *The Open Society and its Enemies* is that it reveals the roots of social and political systems in the allegedly abstract ideas of philosophical thought. Philosophy, far from being considered the irrelevant pastime of an idle or intellectual elite, should for this reason rank as one of the most important disciplines of learning.

Despite the cry to 'keep politics out of religion', politics and religion are inextricably related. The great world religions of Judaism, Christianity and Islam, for example, have been a significant factor in shaping the lives, not only of their adherents, but of the nations where these religions are dominant. One has only to turn to the source book of these three religions, namely the Old Testament, to see the way in which religious and political themes are intertwined. This should not surprise or dismay us, for as we have observed religion is concerned very much with relationships – between man and God; and politics is synonymous with the affairs of men with each other. The plea to keep politics out of religion is understandable when politics has become identified with such evils of irrationalism as sophistry and illusion, with slanging matches and slogans, with bombs and bloodshed. But this makes even more urgent the task of bringing political questions into the public arena of rational debate; and Popper, who has attempted to expose the roots of rationalism, provides a masterly example of how the debate should be conducted. The canons of rational criticism applicable to scientific and metaphysical theories are equally applicable to social and political theories, as Popper demonstrates with a wealth of power and argumentation. This view is supported by one of the most stringent intellectual critics of our time, Isaiah Berlin, who maintains that Popper's *Open Society and its Enemies* contains 'the most scrupulous and formidable criticism of the philosophical and historical doctrines of Marxism by any living writer' (quoted in Magee, 1973, p. 9).

It is not our task here to spell out the details of Popper's social and political theory in the same manner as his philosophy of science, but because his social and political philosophy is of a piece with his philosophy of science we will look at some of the specific issues which bear directly on religious thinking, particularly those associated with *authoritarianism* and *historicism*.

In *Unended Quest* (1976a, Section 24), Popper relates the circumstances under which he wrote his main works on political and social theory. *The Open Society and its Enemies* and *The Poverty of Historicism*. Although they both grew out of the theory of knowledge of *Logic of Scientific Discovery*, the urgency with which they were written was occasioned by events in Europe dominated by Hitler and Stalin and the threat they posed to freedom with their Fascist and totalitarian programmes respectively. Although in New Zealand, far from the scene of conflict, Popper considered these two works as his war effort.

It is not possible even to summarise the prolific weight of argument contained in these books, but the ideas they contain provide both a stimulus and challenge to any theology which attempts to embrace the enduring problems of human freedom, tolerance, suffering, fear and above all the relationship between authority and individual responsibility. What is of fundamental importance for the theologian is Popper's insistence that a rational discussion of such issues is dependent on, among other things, one's philosophy of science.

Consistent with Popper's view that problem solving is the mainspring of life is his conception of how society should be organised if it is to be effective and efficient at solving its problems. If rational criticism is the key to the growth of knowledge and the best way of approaching the truth, so rational criticism is the essential feature of what Popper calls the 'open society'. The open society is characterised by the freedom of its members to criticise and reject the ruling government, whereas the 'closed society' is an authoritarian regime which, because it prohibits criticism and change, is anti-rational and will be unable to solve its problems.

This leads Popper to examine the classic twins of political theory, democracy and totalitarianism. Instead of asking, as Plato and many of his successors in political theory did, the authoritarian question 'Who should rule?' the more important question is 'How can we minimise mis-rule?' In other words, 'How can we so organise political institutions that even bad or incompetent rulers can be prevented from doing too much damage?' As in science, so in politics, justification of authority must be replaced by criticism and a weeding out of error. A society which can elect and dismiss its government without the use of force he calls a *democracy*; whereas a society in which the government can only be removed by the force of revolution is a *tyranny* or *dictatorship*.

There is no doubt that the appeal to authority in science and in society has roots that go far back into man's history and deep into his psyche. With the emergence of man from closed, authoritarian, tribal societies and the beginning of the critical tradition new and frightening demands began to be made which created tensions that haunt man to this day. The conflict between man's instinct to return to the womb of authoritarian security and the awakening of individual responsibility has characterised the political and religious life of man throughout his history. Many would claim that this is the key to the understanding of the Bible in which the story of man is told in terms of his coming to terms with his freedom and responsibility as a son of God. Certainly those who so interpret the source book of the Christian Church would do well to examine Popper's arguments with regard to freedom and authority in political and social life.

If Plato is depicted by Popper as the most influential example of a philosopher of genius whose political theory is inspired by a wish to return to the security of the past, Marx is presented as the supreme thinker of our time whose theory predicts a future resolved of all conflicts. Such reactionary and Utopian theories do not necessarily arise out of human wickedness, but usually out of deep moral concerns for a better world. This is as true of Marx as it is of Plato, and Popper recognises the force of Marx's account of the appalling injustices of nineteenth-century industrial Britain when he writes, in praise of Marx's 'moral radicalism', 'his burning protest against these crimes, which were then tolerated, and sometimes even defended, not only by professional economists but also by churchmen, will secure him forever a place among the liberators of mankind' (1966, Vol. 2, p. 122). The fact that one can often attribute the highest motives to such 'idealists' should not divert our attention from a critical examination of their theories and their consequences. Sincerity must not be confused with truth nor moral rectitude with what is good.

In Popper's view, the best society, in terms of practical and moral considerations, is the open society in which free institutions exist which allow for the removal of the government by the governed, and in which tolerance and freedom are qualified in order to prevent the weak from being exploited and denied their freedom. Force is morally justified only against those who succeed in rendering the free institutions ineffective and attempt to impose authoritarian government.

If what Popper says about the dangers of authoritarian societies has any truth, then his arguments will have some bearing on the problem of authority in religious institutions. The fundamental divisions in the Christian Church, for instance, which have existed since at least the eleventh century, result not so much from differences in interpretation of the deposit of faith, but from differences as to where authority should reside. At one extreme there are those who appeal to the absolute authority of the individual conscience, whilst at the other there are those who insist that the final authority resides in the institution or in the divinely appointed leader. If ever this question is to be resolved, then Popper's discussion of authority resulting from his theory of knowledge and his theory of society is highly relevant.

Perhaps the most important component of Popper's social and political theory is to be located in his discussion of what he calls 'historicism' and his explanation of human action.

> I mean by 'historicism' an approach to the social sciences which assumes that *historical prediction* is their principle aim, and which assumes that this aim is attainable by discovering the 'rhythms' or the 'patterns', the 'laws' or the 'trends' that underlie the evolution of history. (1957, p. 3)

The idea of historical inevitability, that the procession of history is like the procession of the planets, is the basis of many myths and beliefs, some with sinister implications and disastrous consequences. It has inspired people to believe in the idea of the master race, the inevitability of socialism, the perfectibility of man, the Chosen People of the Old Testament and the Second Coming of the New Testament.

Popper's attack on historicist ideas, especially that of Marxism, is formidable, and once again his criticism is rooted in his philosophy of science. He distinguishes between the sorts of prediction scientists make, say about the movement of the tides or the planets, and predictions that are consequent upon experimentation, say in physics. Such experimentation depends upon human action: 'if you do x, then y will follow'. The social sciences, he maintains, involve the latter attempt at prediction and are not in the business of discovering laws of progress or any other laws. His point is that history is made by people, and not directed by some extra-human force, or by the irresitible forces of materialism. Amongst his many arguments against historical determinism is his claim that it is logically impossible to predict future knowledge. 'For he who could so predict today by scientific means our discoveries of tomorrow could make them today; which would mean that there would be an end to the growth of knowledge' (1979, p. 298).

Popper's argument is that history has meaning because of the actions of human beings. Historical and social change results from our attempts to solve problems, and our attempts have both intended and unintended consequences. Central to Popper's argument is his emphasis on individual human responsibility, and it is often our unwillingness to admit to, or to exercise, this responsibility that leads people to seek the false security of authoritarianism. The interplay between individual responsibility and the institutions of which one is a part is a complex issue, and some, like Winch (1958), explain individual behaviour in terms of a Wittgensteinian Form of Life. But Popper's central thesis, that it is individuals that make history, and not history that makes individuals, is a powerful one that has not been refuted.

The conflict between the historicist claim that history is somehow pre-ordained and the indeterminist argument that rests on the notion of human responsibility is apparent in the unresolved tension in theology between the idea that 'God is the Lord of history' and the freedom of the individual to shape his destiny.

Popper's main target in his attack on historicism is Marxism. Once again, the roots of this extraordinarily powerful and appealing political theory are found in natural science. Marx, like so many of his contemporaries, was impressed by the claims of nineteenth-century science, and just as he believed the task of science to be the discovery of the natural laws controlling the universe, so he set out to reveal the inexorable laws of history. Here we must distinguish between Marxism and latter-day communism which is more in the nature of Utopianism. Marxism, as a scientific theory, has testable consequences, which because they have been falsified, is refuted. But not only does Popper demolish Marxist claims to scientific truth, he also reveals the incoherence and inconsistency of Marxist explanatory theory. Magee's conclusion is that 'I do not see how any rational man can have read Popper's critique of Marx and still be a Marxist' (Magee, 1973, p.92). Remarking on the phenomenal influence of Marxism in the world today (one-third of the entire human race have adopted forms of society which call themselves Marxist) Magee continues that 'an understanding of the world we live in today is impossible without some knowledge of his political and social thought' (p. 94).[22] If Magee is right, then the Christian, for instance, who attempts to arrive at an understanding of the world he lives in must have a critical awareness of the source of this most pervasive and influential of political theories.

Derivatives of historicism are Utopianism and what Popper calls 'holism'. Holistic ideas are at the heart of Utopian attempts to reform the whole of society at one sweep. These ideas are, in Popper's view, based on the false assumption that society can be treated as a single entity which can be shaped and manipulated according to some ideal blueprint. This leads to authoritarianism, intolerance and violence, because once the end or goal is defined, no one is allowed to criticise or attempt to change it. Wholesale planning of the Utopian kind does not take into account the unintended and therefore unpredictable consequences of human actions, and in order to contain them it becomes oppressive and violent.

In his essay 'Utopia and Violence' (1972, Ch. 18), Popper demonstrates with devastating clarity the consequences of wholesale Utopian reform. Although written over thirty years before, it is an accurate and fearful picture of Iran in the aftermath of the 1980 revolution which was inspired by Muslim religious fanaticism.

Instead of wholesale Utopian reconstructions of society, Popper advocates what he calls 'piecemeal social engineering' which involves step-by-step changes and improvements. The holistic Utilitarian aim of attempting to achieve the maximum happiness for the maximum number is impossible to achieve because we simply do not know how to make people happy. But we do know sometimes how to remove unhappiness. The logical asymmetry between happiness and unhappiness is like that between verification and falsification in science. Thus we should adopt the more realisable aim of attempting to minimise avoidable suffering, and instead of planning an imaginary Utopia, try to remove the evils before our eyes.

Popper's warnings against Utopianism and holistic reform need to be taken to heart by the man who reckons that wholesale religious reform will save the world. It is one thing to have aims and ideals and to allow them to guide one's actions, but the vision of a new society, of the

Kingdom of Heaven on earth, *can* lead to holistic remedies that end up by enslaving rather than freeing people.

> Of all political ideals, that of making the people happy is perhaps the most dangerous one. It leads invariably to the attempt to impose our scale of 'higher' values upon others, in order to make them realise what seems to us of greatest importance for their happiness; in order as it were, to save their souls. It leads to Utopianism and Romanticism. We all feel certain that everybody would be happy in the beautiful, the perfect community of our dreams. And no doubt, there would be heaven on earth if we could all love one another. But . . . the attempt to make heaven on earth invariably produces hell. It leads to intolerance. It leads to religious wars, and to the saving of souls through the Inquisition. (1966, Vol. 2, p. 237)

Popper sums up what he considers to be a rational assessment of our moral duties:

> It is our duty to help those who need our help: but it cannot be our duty to make others happy, since this does not depend on us, and since it would only too often mean intruding on the privacy of those towards whom we have such amiable intentions. The political demand for piecemeal (as opposed to Utopian) methods corresponds to the decision that the fight against suffering must be considered a duty, while the right to care for the happiness of others might be considered a privilege confined to the close circle of their friends. In their case we may perhaps have a certain right to impose our scale of values — our preferences regarding music, for example. This right of ours exists only if, and because, they can get rid of us; because friendships can be ended. But the use of political means for imposing our scale of values upon others is a very different matter. Pain, suffering, injustice, and their prevention, these are the eternal problems of public morals, the 'agenda' of public policy . . . The 'higher' values should be considered as 'non-agenda' and should be left to the realm of *laissez-faire*. (1966, Vol. 2, p. 237)

Popper's rational approach to social and political values, far from being cold and detached, displays a sensitivity and depth of spirituality that many who pontificate about 'Christian ethics' would do well to emulate.

The Open Universe[23]

In the preface to *The Open Universe* Popper writes:

> I want to state clearly here something that is apparent both in *The Open Society and its Enemies* and in *The Poverty of Historicism*: that I am deeply interested in the philosophical defence of human freedom, of human creativity, and of what is traditionally called free will . . . This book is then a kind of prolegomenon to the question of human freedom and creativity, and makes room for it physically and cosmologically in a way that does not depend on verbal analyses. (p. xxi)

Popper observes that much of the philosophical discussion about freedom of choice and action has been largely verbal, depending on the verbal analysis of the meaning of such words as 'free', 'will' and 'action', and in his strongly anti-essentialist vein insists that 'These verbal analyses are quite futile and have led modern philosophy into a morass' (p. xx).

As usual Popper's avoidance of Essentialism leads him into the heart of the *problem* posed by determinism and which received its strongest formulation by Laplace who maintained that the state of the universe at any time is completely determined and therefore, in principle, predictable. This 'scientific', Laplacean determinism which was an unquestioned part of the Received View of science is 'the most solid and serious difficulty in the way of an account of, and a defence of, human freedom, creativity, and responsibility' (p. xx). Despite the ingenuity and subtlety of linguistic analysts to establish the sense of freedom of action in a physically determined world, Popper perceives that 'if nature were fully deterministic then so would be the realm of human actions; in fact there would be no actions, but at most the appearance of actions' (p. 127). In order then to assess the case for human freedom, Popper sets out to examine the validity of the case for scientific determinism, which he summarises as

> the doctrine that the structure of the world is such that *any event can be rationally predicted, with any desired degree of precision, if we are given a sufficiently precise description of past events, together with all the laws of nature*. (pp. 1-2)

This is the origin of 'Laplace's demon', a fictional, superhuman intelligence who could predict with absolute precision all future states of the world. Popper sums up the intuitive idea behind this rigid determinism by likening the world to a motion-picture film: 'the picture or still which is just being projected is *the present*. Those parts of the film which have already been shown constitute *the past*. And those which have not yet been shown constitute *the future*' (p. 5).

The idea of determinism has its roots in religion and is connected with the ideas of divine omnipotence and divine omniscience, according to which God has complete knowledge of the future which he has determined in advance. The appeal of this deterministic conception of the divine power relates to the appeal of authority which relieves man of the awful burden of personal responsibility for his destiny. Although Luther and Calvin, supported by arguments from St Augustine, developed a brand of determinism, Christian theology has, by and large, adopted the indeterminist position in order to defend free will and so allow man's free response to God's love.

What Popper's thesis demonstrates is that the arguments for free will, which have traditionally been the preserve of religious apologists, are rooted in the philosophy of science. The nettle of *scientific* determinism must be firmly grasped before we can begin to talk about *human* freedom. We have here yet another example of the primacy of an understanding of the philosophy of science for the philosophy of religion. Popper has produced the finest case for freedom, creativity and moral responsibility (all essential items on the agenda of the religious apologist) from considerations arising from the philosophy of science.

Linked with religious and scientific determinism is what Popper labels *metaphysical determinism*.

> The metaphysical doctrine of determinism simply asserts that all events in this world are fixed, or unalterable, or pre-determined. It does not assert that they are known to anybody, or predictable by scientific means. But it asserts that the future is as little changeable as is the past. (p. 8)

Clearly this extreme metaphysical thesis is both untestable and irrefutable. It is like the thesis of *philosophical reductionism* (with which it is linked), a philosophical article of faith, an explanation to end all explanations. However, although metaphysical deteminism is irrefutable, its main *raison d'être* is derived from scientific determinism. If scientific determinism is refuted, then the ground is cut from under the thesis of metaphysical determinism.

The same may be said with regard to attempts to account for human behaviour deterministically. Kant, following Hobbes, Spinoza and Hume, advanced the thesis that full psychological information would enable us to 'calculate in advance, and with certainty, as we do with a lunar or solar eclipse, any man's future behaviour' (p. 20). Such arguments are not the result of precise predictions concerning behaviour, they too flow from the prior conviction that the *physical world* is deterministic. There can be no room in a deterministic world for indeterministic behaviour.

Popper reckons that the thesis of determinism is counter-intuitive and therefore the burden of proof lies with the determinist. Nevertheless he presents a whole weight of argument against scientific determinism which he considers to be a conclusive refutation of the thesis. Among the most appealing of the counter-intuitive arguments is the denial of the possibility of creativity. 'The creation of a new work, such as Mozart's G minor Symphony, cannot be predicted, in all its details, by a physicist, or physiologist, who studies in detail Mozart's body — especially his brain — and his physical environment' (p. 41).

The most fundamental philosophical argument against determinism arises out of Popper's philosophy of science which emphasises the approximate character of all scientific knowledge. 'I see our scientific theories as human inventions — nets designed by us to catch the world,' but they will never be perfect instruments for this purpose: our theories reflect our fallibility (p. 42). The determinist confuses precise prediction with the idea of universal causation of classical physics. Granted that we look for causes of events, and even if we admit (contra quantum physics) that every event has a cause, it does not follow logically that every event can be predicted with absolute precision, which is what the determinist requires. Popper argues this point at considerable length in 'Of Clouds and Clocks' (1979, Ch. 6). The physical world consists of systems which at one end of the scale appear to be like the predictable clockwork of the solar system, and at the other end amorphous systems like random clouds of gas particles or a host of locusts. The determinist maintains that if we knew more about the finer details of the 'cloud' structure we could predict its behaviour with clockwork precision. But by the same token, Popper maintains, the more we expose the 'clock' system to microscopic enquiry, the more it appears to be like an unpredictable cloud. Thus predictability is a matter of degree. The finer the structure, the more difficult the account. The indeterminist does not deny that much is predictable, within limits, in the physical world. His point is that there is much that is not predictable, and this is what the determinist cannot allow. Determinism is based very much on the Newtonian view of the solar system but not only has Newtonian mechanics been shown to be an approximation, the solar system itself is an unrepresentative bit of the universe.

A second important argument concerns the asymmetry between past and future. The past is closed and determined by what has happened, but the future is open. The openness of the future is required by Einsteinian Special Relativity. Unlike Laplace's demon the demon of Special Relativity cannot predict, he can only retrodict (see Section 19). The third argument developed by Popper concerns his anti-historicist thesis. We cannot predict our future knowledge; if we could then the knowledge of tomorrow we would have today, and there would be no growth of knowledge. This is linked with Popper's last and most decisive argument concerning the impossibility of self-prediction. We cannot predict our own future states. Lucas has presented a similar argument in his *The Freedom of the Will* (1970), which uses Godel's proof that no calculator can predict the results of its own calculations or predictions.[24] He concludes that, at the human level 'I, and only I, am responsible for my choice. The buck stops here' (Lucas, 1970, p. 172). It is Popper's contention that determinism has no explanatory power. It does not help us to understand the incredibly complex world in which we find ourselves. Indeterminism, on the other hand, has considerable explanatory power, and it produces 'positive gains for common sense, ethics, the philosophy of science, cosmology and, I hope, for truth' (p. 93). And, we would add, for the philosophy of religion.

However, Popper maintains that *indeterminism is not enough* to account for human freedom and creativity. 'For what we want to understand is not only how we may act *unpredictably and in a chance-like* fashion, but how we can act *deliberately and rationally*' (p. 126). The view of Laplace leads to predestination.

It leads to the view that billions of years ago, the elementary particles of World 1 contained the poetry of Homer, the philosophy of Plato, and the symphonies of Beethoven as a seed contains a plant; that human history is predestined, and with it all acts of human creativity. (pp. 127-8)

But to attribute such creativity to sheer chance is also an absurdity. Popper offers an explanation of creativity in terms of his three-world doctrine. World 1 (the world of material things, which is partly but not completely determined) is influenced by World 2 (the world of the human mind), which in turn interacts with World 3 (the world of objective ideas). Determinism insists that World 1 is a closed system, uninfluenced by World 2 which is at most an epiphenomenon. But in order to account for the creative genius of Michelangelo, Mozart or Einstein, or the impact of Karl Marx, or the more commonplace creativity of John Smith, World 3 must influence World 2 and World 1. A degree of 'plastic control' and feedback is involved in the interaction of the three worlds. The universe is bound to be open if it contains human knowledge, for human knowledge is ever incomplete and in constant interaction with World 1. Who can doubt this after Hiroshima?

> No good reason has been offered so far against the openness of our universe, or against the fact that radically new things are constantly emerging from it; and no good reasons have been offered so far that shed doubt upon human freedom and creativity, a creativity which is restricted as well as inspired by the inner structure of World 3.
>
> Indeterminism is not enough: to understand human freedom we need more; we need the openness of World 1 towards World 2, and of World 2 towards World 3, and the autonomous and intrinsic openness of World 3, the world of the products of the human mind and, especially, of human knowledge. (p. 130)

J.W.N. Watkins, in 'The Unity of Popper's Thought' (1974), argues that indeterminism is the most striking feature of Popper's metaphysical outlook and that it is this that gives a unity and coherence to his wide-ranging ideas. He suggests that determinism and inductivism, although not logically related, are mutually supporting theses. They rest on assumptions about the nature of the world and man's relation to it: that all phenomena must ultimately be reduced to physical descriptions, to the realm of empirical science which sets the bounds of rationality. Not only are human actions determined, but the whole universe is one vast determinate system. This thesis, like Lamarckian induction, holds that it is the environment that conditions a person.

There is likewise a natural coalition between indeterminism and falsi-ficationism (whereby scientific knowledge is seen as growing through conjecture and refutation). According to this view, not only is human behaviour ultimately unpredictable (although it is affected by causal factors), but the course of the world is fundamentally indeterminate. Thus man is, in the last analysis, a free and responsible agent whose decisions help to shape the course of events that constitute the history and the future of the world. There is no doubt as to which thesis is most plausible and which offers the more optimistic view: in J.W.N. Watkins' words:

> The first depicts man as an induction machine nudged along by external pressures, and deprived of all initiative and spontaneity. The second gives him the *Spielraum* to originate ideas and try them out. Learning about the world means, on the first view, being conditioned by it; on the second view, it means adventuring within it. (Watkins, 1974, p. 407)

Evolution

In the debate between Bishop Wilberforce of Oxford and Thomas Huxley that followed the publication of Darwin's *The Origin of Species*, Wilberforce wrote that Christianity 'was utterly irreconcilable with the degrading notion of the brute origin of him who was created in the image of God'. The bishop was not so much concerned about the challenge Darwin appeared to present to the authority of the Scriptures, to the literal truth of the Genesis account of creation, but about the dignity of man which he saw threatened by the evolutionary account. The wife of the Bishop of Worcester, with an eye to diplomacy, commented, 'Descended from the apes! My dear let us hope that it is not true, but if it is, let us pray that it may not become generally known.'

However, despite the resistance of such contemporary Establishment figures (scientists included), and the anti-evolutionist stand made by Biblical fundamentalists, the Christian faith has had little difficulty in assimilating evolutionary theory. But although Christian theology has succeeded in syncretising the naturalistic account with God's creativity, it has achieved this almost, as it were, by a theological sleight of hand; by superimposing God as the originator and sustainer of a naturalistic process. God created the universe in such a way that life would arise and evolve (more than a hint of determinism here); it is still God's creation. Consequently evolution is depicted as an isolated biological theory, another scientific theory of which theology takes note, but which it does not meet at any deep level of understanding.

The significance of Popper's evolutionary philosophy is that it sets biological evolution in a much wider metaphysical context, which has a remarkable degree of inner coherence and logical consistency, and great explanatory power. And far from degrading man and robbing him of his unique dignity, Popper sees man as the great miracle of evolution.

Where evolution is concerned, Popper is not sparing in his use of the term 'miracle', with all its overtones of inexplicability, and sense of breathtaking awe and wonder. The three great 'miracles' of life are its mysterious origin, the emergence of consciousness, and the emergence of the human mind. Although man is an animal, and stands nearer to his fellow animals than they stand to inanimate matter, we should not 'belittle the gulf which separates the human brain from the animal brain, and human language from all other animal languages' (1982b, p. 122).

In the preface to *The Self and its Brain* Popper and Eccles admit that one of their motives for writing the book is that

> we both feel that the debunking of man has gone far enough — even too far. It is said that we had to learn from Copernicus and Darwin that man's place in the universe is not so exalted or so exclusive as man once thought. That may well be. But since Copernicus we have learned to appreciate how wonderful and rare, perhaps even unique, our little Earth is in this big universe; and since Darwin we have learned more about the incredible organization of all living things on Earth, and also about the unique position of man among his fellow creatures. (1977, Preface)

The human self is the most remarkable and unlikely known phenomenon in the universe. The mystery of Mozart is not reduced by his animal antecedents: it is enhanced and enriched. This sense of awe and wonder, and mystery that characterises Popper's conception of science is precisely that sense of awe, wonder and mystery that characterises the religious response.

We will conclude by examining a recent exercise in theological method that embraces the points we have so far made concerning the rational argument of metaphysical theses, the openness of theology to new ideas, and above all the evolutionary thesis with its implications for the open universe.

Of increasing importance in Popper's later writings is the idea of human creativity, and his attempt to account for it in terms of the openness of, and interaction between, Worlds 1, 2 and 3. W.H. Vanstone, in a remarkable book, *Love's Endeavour, Love's Expense* (1977), offers an account of the creative love of God which has truly revolutionary implications for what we might call the 'received view' of theology.

According to this 'received view', and reflected in popular devotion,

God is conceived primarily in terms of power. The power of God is an extrapolation of earthly power: a power which, since Hiroshima, appears to be ever increasing in the hands of man. Vanstone points to the moral ambiguity of power: power involves privilege, not worth. Superiority of power confers no right to respect. The integrity of religion is at stake here, and modern man has become increasingly aware of the hypocrisy implicit in worship which is based on the adulation of power.[25] Weakness, on the other hand, does make profound moral claims upon us, and it is doubtless for this reason that the contemporary Church has become increasingly identified with the underprivileged and helpless and oppressed. This sensitivity to helplessness and weakness derives from the recovery of the New Testament idea of *kenosis*, or *self-emptying*; of the power of God in weakness. The claim of God on us is not the authoritarian claim of power, but the moral claim of weakness. This led Vanstone to examine afresh the idea of 'the love of God', and in particular what is involved in God's creative love.

Our notion of 'the love of God' derives from our experience of human love. And although human love is ever imperfect, we recognise from its imperfection what authentic love should be like. (Here is an example of a theological idea being tested against experience.) Our experience of human love leads us to an approximation of God's love as *limitless*, *precarious* and *vulnerable*. At this point Vanstone is, in a sense, effecting a revolution in our thinking about the creative love of God, and one which has remarkably fruitful consequences for our understanding of God and his relationship with the world.

The activity of God in creation must be *limitless*. We are not to imagine a God who, as it were, creates with his left hand the being of the universe, while reserving to himself unexpended stores of power and potentiality. If God is love, God is wholly expended in his creative activity.

> Nothing must be withheld from the self-giving which is creation: no unexpended reserves of divine power or potentiality: no 'glory of God' or 'majesty of God' which may be compared and contrasted with the glory of the galaxies and the majesty of the universe: no 'power of God' which might exceed and over-ride the God-given powers of the universe: no 'eternity of God' which might outlive an eternal universe. It is to be understood that the universe is not to be equated with 'that which science knows,' nor even with 'that which science might, in principle, come to know': the universe is the

totality of being for which God gives Himself in love. From His self-giving nothing is held back: nothing remains in God unexpended. (Vanstone, 1977, pp. 59-60)

The activity of God in creation must be *precarious*. It must proceed by no assured programme. Each step is a precarious step into the unknown, involving triumph or tragedy. God does not work to fore-ordained programmes and purposes. Productivity is not creativity.

If the purpose of God in creation is foreknown and foreordained to fulfilment, then the creation itself is vanity. Within it nothing decisive happens, and nothing new: it is merely the unwinding and display of a film already made. On the other hand, to interpret the creation as the work of love is to interpret it as the new, the coming-to-be of the hitherto unknown, and so as that for which there can be no precedent and no programme. If the creation is the work of love, its 'security' lies not in its conformity to some predetermined plan but in the unsparing love which will not abandon a single fragment of it, and man's assurance must be the assurance not that all that happens is determined by God's plan but that all that happens is encompassed by His love. (Vanstone, 1977, pp. 65-6)

The activity of God in creation must be *vulnerable*. The lover cannot remain indifferent to the object of his love: he becomes vulnerable to the response of the beloved. Likewise God's creative love involves his vulnerability, or susceptibility. Traditionally God has been conceived as impassible, unexposed or unaffected by what happens in the world. If God suffers with the world, how can we avoid pantheism or anthropomorphism? Vanstone considers this problem at greater length in a subsequent book *The Stature of Waiting* (1982), in which he interprets God's impassibility in terms of waiting: waiting for the response of his beloved creation.

Vanstone is critical of the conventional representation of God as 'One by Whom, in the creation, nothing is expended and nothing jeopardised, who presides serene over the assured unfolding of a predetermined purpose'. Rather:

If God is love, and if the universe is His creation, then for the being of the universe God is totally expended in precarious endeavour, of which the issue, as triumph or as tragedy, has passed from His hands. For that issue, as triumphant or as tragic, God waits upon the response of His creation. He waits as the artist or as the lover waits, having given all. Where the issue is tragedy, there remains only the unbelievable power of art or love to discover within itself, through the challenge of the tragic, the power which was not there before — the power of yet further endeavour to win back and redeem that which was going astray. Where the issue is triumph, there remains only the will of love to surrender triumphant self-sufficiency in yet larger, more distant, more generous endeavour. Always, for the richness of the creation, God is made poor: and for its fulness God is made empty. Always His helplessness waits upon the response of the creation. To anyone who does not understand this, or cannot accept this, we must answer, 'Nondum considerasti quanti ponderis sit amor, quanti ponderis creatio.' ('You have not yet weighed the cost of love, the cost of creation.') (Vanstone, 1977, p. 74)

Vanstone proceeds to examine three levels of response to God's creative love: the response of nature; the response of freedom; and the response of recognition. *The response of nature* takes the form of actual material things, from stones to stars. If the creation matters it matters *now*, for what it is, not for some future realisation or consummation 'to which the whole creation moves'. Here is an echo of Popper's anti-historicism, a denial of the evolutionary and progressivistic thinking which characterised an influential but illegitimate interpretation of nineteenth-century Darwinism.

There is no reason to believe in a future which will contain the key to the meaning and purpose of the present. If there is no purpose to be detected in the concrete and contemporary reality of a particular stone or tree or star, then no purpose will be detected in the total reality of the universe a thousand million years from now. (Vanstone, 1977, p. 80)

The response of freedom involves free human choice and the performance of concrete material actions. It consists of the good deeds of good men in response to God's creative love. Again, this is in accordance with Popper's thought concerning human freedom and responsibility, especially with its emphasis on the importance of our response to 'the concrete and individual crises of human existence' (Vanstone, 1977, p. 92)

The response of recognition is the free response to God's creative love, publicly expressed. It is the *celebration* of the love of God. 'The final triumph of the love of God is the celebration of His love within that universe which has received that love' (Vanstone, 1977, p. 97). This, Vanstone maintains, is the *raison d'être* of the Church: it occupies the enclave of recognition within the area of freedom. And that is why *worship* is central to all religion: it is a recognition of the love of God expended and exhausted to the point of weakness.

Thus Vanstone demonstrates how even such a fundamental and crucial issue as 'the love of God' can be open to fresh analysis and developed with fresh insight, tested against experience. It also provides a coherent theological account of divine creativity which is consistent with biological and cultural evolution with its notions of freedom and indeterminism, and with the risky and precarious nature of natural selection.

We have come full circle in our discussion of Popper's philosophy of science and its relevance for a critical, rational theology. An account of biological and cultural evolution, which is central to the descriptive epistemology which informs Popper's philosophy, is also of direct relevance to our understanding of the creativity of God, which is at the heart of all religious thought.

Postscript

If we have suggested fruitful links between some of the fundamental philosophical questions faced by science and religion in Popperian terms, it does not follow that Popper's philosophy of science is inviolable; but neither does it follow that because there are flaws in the Popperian scheme, the whole edifice comes tumbling down.

We began by attempting to give an account of Popper's descriptive epistemology, and in this matter Popper has made an enormous and lasting contribution to the whole approach to the problem of knowledge. Science and philosophy are inseparable. A purely conceptual account of knowledge is replaced by a view which takes into account the biological structures and the social institutions which characterise man as knower. If this is important for the relationship between science and philosophy, then by the same token it is important for the relationship between philosophy and religion; and it implies that philosophy is the link between the two enterprises.

This raises the question, once again, of demarcation. Can there be a clear, definitive demarcation between science and what Popper calls 'non-science'? Popper's proposal for a logical distinction is clear; it is logical falsification. However, it is the attempt to put this into practice that leads into the troubled waters of methodological falsification. There is no doubt that in practice the line between science and non-science is fuzzy; but that there *is* a distinction between the two is central to Popper's whole philosophy. What Popper has established is that scientific questions can be refuted with greater precision than can the psychoanalytic theories to which he refers. In this respect Popper's view has recently been reinforced by the frustrated attempts of psychiatrists to specify the conditions of mental abnormality in regard to the mass murderers Sutcliffe and Nilsen. Dr Gallway, a defence psychiatrist in the Nilsen trial, offered an account of the defendant's psychological state that was impossible to refute; and his comment that the syndrome was 'best diagnosed by someone with my particular interests' is reminiscent of Adler's reply to Popper about his 'thousandfold experience' (see Nicholson-Lord, 1983).

This in turn brings us back to the question of whether or not Popper has completely laid the ghost of induction. Possibly he has not. We would agree with Lakatos' 'thin metaphysical principle of induction' which allows us to distinguish between constructive fallibilism and profound scepticism. But this may be no more than a verbal quibble, for Popper makes it clear that provisional justifications, or subjective metaphysical beliefs concerning regularities and expectations, are necessary for practical action. Such an admission in no way undermines Popper's methodology; and his suggestions regarding conjecture and refutation have proved most fruitful, as has his instructive distinction between reliance and reliability, between practice and the logic of the situation. His point is that the rational man will ever remain critical of his most cherished assumptions.[26] Neither is the fundamental thesis of indeterminism in any way weakened by the admission of such a tentative link. We agree with J.W.N. Watkins that the thesis of indeterminism is crucial to Popper's metaphysical outlook, and that it offers the most plausible and fruitful account of the physical world of which we are an integral part and in which we are free (within limits), conscious participators.

The most important and difficult issue of all concerns the question of truth. Many people are mystified by Popper's claim that Tarski has rehabilitated the correspondence theory of truth, and his insistence that this is the central idea of his whole philosophy. Once again, Popper's position becomes clearer when we pay attention to his distinction between logic and method. The question remains, does it make sense to speak of the metaphysical ideal of absolute truth? This is indeed a portentous issue on which the greatest of philosophers have foundered, and it is probably one of the central issues in current philosophical debate. However this may be, we do well to heed Popper's dictum that 'truth is hard to come by'. Whether or not it is impossible even to conceive of is a question to which we will now turn.

Notes

1. All references in this chapter are to Popper unless otherwise indicated.

2. Sir Hermann Bondi, Sir John Eccles and Sir Peter Medawar, to mention three 'Popperian Knights', all distinguished scientists in their respective fields, have all admitted to a change of attitude and method as a result of Popper's influence. They agree that 'The wrong view of science betrays itself in the craving to be right' (1968a, p. 281). See for instance Eccles' essay in Schilpp, 1974, Ch. 10.

3. The notable exception appears to be his theory of the three worlds which he develops in *Objective Knowledge* and subsequent writings.

4. J.C. Eccles in *The Human Mystery* (1979) attempts to relate the three phases of evolution using Popperian insights and his own expertise in the field of neurophysiology.

5. Professor Fred Hoyle has also recently questioned Darwinian natural selection in terms of the impossibly large time scale that would be needed for the accumulation of favourable chance mutations necessary for even the simplest organisms. He offers an alternative 'creationist' hypothesis (Hoyle, 1983).

6. The development of the computer, which itself is the product of man's ability to develop highly abstract theories, could well mark a turning point in man's cultural evolution comparable with that of the invention of writing.

7. Evolutionary epistemology has an interesting history, spanning the nineteenth and twentieth centuries. Gonzalo Munévar traces its development from Spencer to Popper and Toulmin (1981, Ch. 6). Our concern is specifically with Popper's account and its bearing on his theory of knowledge in general and his theory of science in particular. Campbell reckons that 'it is primarily through the works of Karl Popper that a natural selection epistemology is available today' (Campbell, 1974, p. 413)

8. *The Postscript to the Logic of Scientific Discovery*, edited and published under three separate titles: Popper, 1982a, 1982b, 1983.

9. See 'Of Clouds and Clocks' in Popper, 1979; and 'Indeterminism is not Enough' in Popper, 1982b.

10. One must distinguish between the theories of Marx and Freud and their dogmatic champions. Certainly the theories of Marx are open to falsification, as Popper himself acknowledges, and many of the theories of Freud too are arguably falsifiable. It is the supporters of Freud who refuse to accept falsifications. Although this reduces to a problem concerning the psychology of those who cling to their theories, come what may, Popper's distinction between all-embracing theories to be confirmed by all possible events and theories open to critical experiential testing remains.

11. See also Einstein's letter to Popper in Popper, 1968a, p. 458.

12. No attempt at formulating inductive logic has met with success. 'Inductive Logic', so called, cannot be successfully programmed on a computer, whereas recent developments in deductive logic coupled with remarkable advances in computer technology present a completely different picture in terms of rigour and proof in the mathematical sciences.

13. In his 1983 introduction to *Realism and the Aim of Science* Popper again attempts to set the record straight regarding *logical* falsification and

methodological falsification. One must not confuse what is possible in principle, or in logic, with what is possible in practice. Popper writes:

> An entire literature rests on the failure to observe this distinction . . . it has led some people to abandon rationalism in the theory of science, and to tumble into irrationalism. For if science does not advance rationally and critically, how can we hope that rational decisions will be made anywhere else? A flippant attack on a misunderstood logical-technical term has thus led some people to far-reaching and disastrous philosophical and even political conclusions. (1983, pp. xxii-xxiii)

14. This is breathtaking nonsense and could only have been written by a philosopher in a hurry. It exaggerates Popper's failure to supply a formal account of verisimilitude out of all proportion to its importance.

15. Lakatos, in 'Falsification and the Methodology of Scientific Research Programmes' (1970), introduces the idea of *Scientific Research Programmes* as a corrective to Popper's falsification methodology without any acknowledgement to Popper, who had already developed the idea (see Editor's note in Popper, 1982a, p. 32). Popper's *Metaphysical Research Programmes* also anticipated, in some respects, Kuhn's *Paradigms* (Popper, 1982a, p. 31). See also Popper, 1976a, pp. 148-51 and note 242.

16. See note 2 in which Popper makes an interesting reference in this matter to 'historicist nonsense'.

17. Popper is mistaken in this comparison. By his own argument the Book of Revelation is open to rational discussion to the extent that it can be shown to 'tie up with something else'; i.e. the rest of the New Testament and the events to which it bears witness.

18. See, for instance, Keith Ward, *Rational Theology and the Creativity of God* (1982b), in which he maintains that 'belief in God is the highest expression of human rationality' (p. 1).

19. *God Has Many Names* (1980), *God and the Universe of Faiths* (1977), 'Jesus and the World Religions' (1977b).

20. Emphasis mine.

21. We discuss this more fully under the heading 'Evolution', pp. 127-32.

22. Clearly, there is no space here for a full discussion of Popper's views on historicism and the critical response of defenders of Marxism, such as Marcuse and members of the Frankfurt School, who have attempted to rehabilitate Marxist theory and who deny that it has been falsified in the manner in which Popper suggests (see, for instance, Marcuse, 1978).

23. All references in this section are to page numbers in *The Open Universe* (Popper, 1982b), unless otherwise stated.

24. Although Kurt Godel was primarily a mathematical logician, his discoveries in the early 1930s rank him as one of the most outstanding philosophers of our time. Various attempts have been made to translate Godel's 'incompleteness theorem' into a general philosophical principle but the most striking generalisation is that given by Rudy Rucker in his recollections of discussions with Godel in his book *Infinity and the Mind* (1982): 'we can never know the absolute truth', a Kantian and a Popperian conclusion that is derived with impeccable logical and mathematical rigour.

25. Popper speaks of 'the adulation of power' in *The Open Universe* (1982b, p. 5). A whole weight of theology is contained in this footnote.

26. There is no doubt that much human suffering could be prevented if inductive assumptions ('it has never happened before') were replaced by a more rigorous and searching critical attitude.

3 THE RETREAT TO IRRATIONALITY

The Rationale of Discovery

One of the curious things about *The Logic of Scientific Discovery* is that it is not about the logic of scientific discovery. Popper's central concern is with the *growth* of scientific knowledge, and he attempts to give an account of the logic of this growth. He denies that there is any specific scientific method or logic of discovery. 'There is no such thing as a logical method of having new ideas, or a logical reconstruction of this process . . . every discovery contains an "irrational element", or a "creative intuition" ' (1968a, p. 32). Although Popper's arguments against the idea of a logic of discovery are derived from his rejection of inductive logic, which suggests that scientific theories can be inferred from sets of observations, he is nevertheless at one with the approach of the later representatives of the Received View in their exclusion of the context of discovery from their rational reconstruction of science.[1]

But despite Popper's attempt to draw a sharp line between discovery and testing, it was his attempt to give a formal account of the testing of scientific theories that opened the way to a new philosophy of science which took into account the rationale of discovery. This arose out of the tension, which is evident throughout Popper's work, between the formal and the practical, which we attempted to expose in the distinction between logical and methodological falsification.

Logical falsification is rooted in the formalism of the *modus tollens* of deductive logic, whilst methodological falsification is sensitive to the theory-dependence of observation, and the complexity of theories in their internal structure and their relationships with other theories. It is because a satisfactory formal account of methodological falsification eluded Popper that the idea of science as a social phenomenon was increasingly exploited.

This shift in attention from analysis of the logical structure of theories to the attempt to understand the rationale of scientific discovery and theory change resulted in the increasing recognition of the importance of the theoretical framework which determines not only the problems to be solved, but the nature of the solutions to these problems. Thus science is done from within a *Weltanschauung*, or conceptual perspective, which is intimately linked with the way our language conceptually shapes our understanding of the world.

The *Weltanschauung* approach attempts to shift the boundary, originally set by the empiricists, between what is and what is not relevant to the philosophical analysis of science, to include history, sociology, psychology, and even the economics and politics of science. The fundamental shift occurred as a result of the admission of the theory-dependence of observation (which was at the heart of the collapse of the Received View) originally perceived by Popper, and followed up by (among others) Polanyi (1957), Hanson (1958), Toulmin (1961), Kuhn (1970a) and Feyerabend (1981). The most influential of the *Weltanschauungen* analyses was that offered by Kuhn, and more recently exploited by Feyerabend; and as we shall indicate, the *Weltanschauung* approach has been thoroughly exploited by some philosophers of religion.

Although it is argued (e.g. by Suppe) that the *Weltanschauung* view arose as a reaction to the failure of logical empiricism to provide a satisfactory account of the structure of scientific theories, both Kuhn and Feyerabend acknowledge their debt to Popper, and initially they saw their ideas as a development of Popperian philosophy of science. Kuhn is at pains to stress the points at which he is in agreement with Popper, especially in his shared concern with the realities of scientific practice, his insistence that knowledge does not grow by accretion, the revolutionary nature of theory rejection, and the interdependence of observation and theory. Thus he writes that at many points 'Sir Karl's view of science and my own are very nearly identical,' and the extensive agreement between them 'is enough to place us in the same minority among contemporary philosophers of science' (Kuhn, 1970b, pp. 1-2). And Feyerabend, before he developed his more extreme views and his antipathy to Popper, acknowledged, 'I for one am not aware of having produced a single idea that is not already contained in the realistic tradition and especially in Professor Popper's account of it' (quoted by Suppe, 1977, p. 166). However, as Kuhn and Feyerabend developed their ideas they increasingly distanced themselves from Popper, particularly over the crucial issues concerning demarcation, truth and realism; and Popper has remained vehement in his denunciation of Kuhn's and Feyerabend's relativism, which he dubs *The Myth of the Framework* (1976a), and in his hostility to the reduction of the philosophy of science to psychology and sociology.[2]

In some respects Kuhn and Feyerabend between them turned many of the assumptions of academic thinking about science upside down; they certainly raised some awkward questions about the orthodox picture of science as an objective, fact-finding, or theory-testing enterprise. The most radical conclusion to be drawn from their work is that scientific facts or theories are made and not discovered. No longer can the scientist be depicted as the detached investigator in a world of facts: he helps to shape the world he investigates. The relativism which many 'rationalists' reckoned was at the heart of the irrational affairs of men in ethics and religion was now invading the sacrosanct bastion of rationality, natural science.

We will begin by outlining the central ideas contained in Kuhn's seminal and influential book, *The Structure of Scientific Revolutions*. Although its author has retreated, under criticism, to a much more conservative position since that book was published, Feyerabend in particular has developed those ideas in a much more radical and challenging manner. Our discussion, however, is more concerned with the ideas and their implications, rather than with the question of whether or not a particular idea is held at a particular time by a particular person.

T.S. Kuhn[3]

'What must nature, including man, be like in order that science be possible at all?' (1970a, p. 173). This is the central question running through *The Structure of Scientific Revolutions*. As with so many problems, the formulation of the question is more important than the answer. Kuhn does not claim to have answered the question, but his attempt to open up the relationship between nature and man ranks as a fundamental and original contribution to the philosophy of science.

Kuhn sets the scene of his investigation with the opening words of *The Structure of Scientific Revolutions*: 'History, if viewed as a repository for more than anecdote or chronology, could produce a decisive transformation in the image of science by which we are now possessed' (1970a, p. 1). The representatives of the Received View tended to view the history of science simply as a repository of knowledge, to which reference could be made to support their logical analyses; whereas Kuhn maintains that it is impossible to separate scientific knowledge and reasoning from its historical development. History, according to Kuhn, does not display a steady accumulation of knowledge as the great men of science increasingly come to grips with the world through theory and experiment. Rather, history displays a variety of traditions or 'paradigms' in which men are bound by shared presuppositions or underlying principles which determine the way they see the world. The replacement of one tradition or 'paradigm' by another is more like a change of conceptual spectacles than a steady progression from the darkness of ignorance and superstition and prejudice into the increasing light cast by objective science. Scientific progress is viewed, not in terms of an end or goal such as increasing verisimilitude, but in terms of change. And if we are to understand scientific change we must examine not merely the logical form of theories, but the actual content of particular scientific views. The theory of knowledge has become the sociology of knowledge.

Thus it is to the realities of scientific practice, rather than the objects of scientific discovery, that Kuhn turns his attention. And what he sees is a dynamically evolving *Weltanschauung*, or a conceptual perspective. The evolution of this *Weltanschauung* is not a smooth procession in which scientific knowledge is accumulated, but rather it is a discontinuous affair punctuated by revolutionary upheaval. Unlike Popper, who is concerned with the *growth of scientific knowledge*, Kuhn is concerned with *scientific change* which, he maintains, is fundamentally revolutionary.[4]

The revolution takes place when the 'paradigms' that characterise non-revolutionary, normal science are overthrown to be replaced by fresh ones. 'Scientific revolutions are . . . those non-cumulative developmental episodes in which an older paradigm is replaced in whole or in part by an incompatible new one' (1970a, p. 92), where 'paradigms' are shared examples of successful practice rather than shared rules of procedure, 'examples which include law, theory, application, and instrumentation together' and which 'provide models from which spring particular coherent traditions of scientific research' (1970a, p. 10).

The stages along the evolutionary path of science can be depicted in the following schema:

pre-science, normal science, crisis, revolution, new normal science

It is an open-ended schema, with no ultimate end or goal, and in which progress is viewed in terms of change. *Pre-science* is a disorganised, diverse activity which becomes structured and directed and community based when a single paradigm, such as Newtonian Mechanics or Relativity theory, is tacitly accepted by the scientific community. *Normal science* assumes the general theories and laws and techniques of the paradigm and involves the scientist in exploiting the paradigm to the fullest extent. His attitude is largely uncritical in that he tends to set aside unsolved puzzles and falsifications in his efforts to provide a satisfactory match between the paradigm and the natural world. A *crisis* occurs when the unsolved problems get out of hand and undermine the

confidence of the normal scientists. It is only resolved when an entirely
new paradigm emerges and attracts the allegiance of the majority of the
scientific community. A *revolution* involves the abandonment of the
old paradigm for the new, and so the community settles down once
again to a welcome period of normal science until such time as *its*
assumptions in turn become so strained that yet a fresh crisis looms.

Paradigms – Disciplinary Matrixes and Exemplars. The plausibility of
Kuhn's thesis is very much dependent on the manner in which he uses
the central concept 'paradigm'. As the book develops, our suspicion
grows that it can be too nearly all things to all men. Indeed, Kuhn has
since admitted that when challenged to explain the absence of an index,
'I regularly point out that its most frequently consulted entry would
be: "Paradigm, 1-172, *passim*"' (1977a, p. 459). It is not surprising
therefore that such an all-embracing term should lead to confusion and
philosophical obscurity, and it at once exposes the inherent weakness
of Kuhn as a philosopher of science. Masterman has discerned at least
twenty-one different uses of the term (1970, p. 61), and Shapere
observes that it includes 'laws and theories, models, standards, and
methods (both theoretical and instrumental), vague intuitions, explicit
or implicit metaphysical beliefs (or prejudices)', and he argues that the
plausibility of Kuhn's thesis is guaranteed by the vague and ambiguous
manner in which the term is used (quoted in Suppe, 1977, p. 137).

Kuhn has since had 'Second Thoughts on Paradigms' (1977a), and he
has attempted to clarify the notion by distinguishing between
exemplars and *disciplinary matrixes* as the two main constituents of
his 'paradigm'. *Exemplars* are concrete problem situations accepted
within the scientific community; and *disciplinary matrixes* are the more
generally shared elements which characterise the scientific community,
such as symbolic generalisations, models, values and metaphysical principles. The disciplinary matrix is a kind of scientific *Weltanschauung*, a
broad conceptual framework, which, because of its all-embracing nature,
cannot be precisely specified. It is not acquired through the study of
explicitly formulated methodological rules, but through tackling the
specific problems, or exemplars, which are the shared concern of the
scientific community. It is through the study of exemplars that one
becomes initiated into the scientific community, by learning the
common scientific language and the sort of questions and answers that
are acceptable within that community. Kuhn is here describing the
familiar process by which a student, through studying specific
examples, say in physics, chemistry or biology, doing experiments,

reading the journals, and eventually pursuing his own research, gains admission into the scientific community. Essential to this process is the interplay between the disciplinary matrix and the exemplars: the scientist obtains his disciplinary matrix through acquaintance with exemplars and they in turn determine the disciplinary matrix. Thus Kuhn writes, 'one sense of "paradigm" is global, embracing all the shared commitments of a scientific group; the other isolates a particularly important sort of commitment and is thus a subset of the first' (1977a, p. 460). Clearly the exemplar is the more fundamental and philosophically more important component of the two for without exemplars one could not obtain a disciplinary matrix.

The interplay between the exemplar and the disciplinary matrix is further complicated by the symbolic generalisations of a theory. It was the failure to reduce successfully the symbolic generalisations of a theory, by correspondence rules, to empirical phenomena, that led to the collapse of the Received View. Kuhn's thesis is that the symbolic generalisations of a theory cannot be explicitly interpreted but rather implicitly assimilated by handling exemplars. Thus the meaning of the terms in a theory is determined by the shared stock of exemplars which in turn determines the disciplinary matrix or view of the world. It follows that different scientific communities possessing different exemplars will attach different meanings to the same theoretical terms and will therefore entertain different theories and will view the world differently. For example, although Newton and Einstein each employed the same term 'mass', they interpreted it in entirely different senses and produced theories which picture the world in entirely different ways. Furthermore, since the exemplars indicate the sorts of questions and answers that are appropriate to the community sharing those exemplars, there will be disagreement between communities with different shared exemplars concerning the relevant questions and answers. That is, the shared exemplars will determine the scientific values adopted by a scientific community. The import of this sociological analysis is that not only can there be no neutral observation language common to all scientific investigators, but there are no objective 'facts' because they too are determined by the exemplars.

Normal Science. Having attempted to clarify and give content to Kuhn's original concept of 'paradigm', we must turn our attention to the dynamics of scientific change which, Kuhn maintains, characterises the history of science. If scientific change is fundamentally revolutionary there must be contrasting periods of normal, non-revolutionary science. It is this *normal science* which characterises the larger part of the history of science and which is 'an enterprise which accounts for the overwhelming majority of the work done in basic science' (1970b, p. 4).

If the bulk of scientific endeavour is located in normal science then its characterisation is of central importance to Kuhn's thesis. Although the title of Kuhn's book implies that scientific revolution is his theme, he makes it clear that normal, non-revolutionary science is, and ought to be, the main part of scientific endeavour. A scientific community bound by a common disciplinary matrix, which in turn is based on a shared stock of exemplars, practises normal science. The activities of the normal scientist all take place within the bounds of the framework supplied by the disciplinary matrix. His task is to co-operate with his fellow scientists in articulating and extending the disciplinary matrix through the production of additional exemplars, or concrete problem situations, and the refinement of the existing ones. Thus Kuhn characterises normal scientific research as 'a strenuous and devoted attempt to force nature into the conceptual boxes supplied by professional education' (1970a, p. 5).

An example of a disciplinary matrix which provides a framework within which normal science takes place is Newtonian Mechanics, with its laws of motion, the application of these laws to all physical phenomena from ideal gas molecules to the solar system, instruments from spring balances to telescopes, and a general metaphysical principle according to which the world is viewed as a mechanical system subject to universal forces. Classical electromagnetic theory, with its notions of force and field and the all-pervading aether, and the elegant symbolism of Maxwell's field equations, constitute another highly fruitful disciplinary matrix, or scientific *Weltanschauung*. Presumably Quantum Mechanics provides a successful example of shared practice which, for all its question marks, still unites contemporary scientists in a disciplinary matrix.

Within the disciplinary matrix the golden rule is not to rock the boat with criticism, but endeavour to match the theory with nature. The aim is not novelty or surprise but conformity to the basic world view supplied by the disciplinary matrix. The activity of the normal scientist is not so much testing in the Popperian sense, but what Kuhn calls *puzzle solving*. 'To rely on testing as the mark of a science is to miss what scientists mostly do and, with it, the most characteristic feature of their enterprise' (1970b, p. 10) and, later, 'of the two criteria, testing and puzzle solving, the latter is at once the less equivocal and the more fundamental' (1970b, p. 7). Thus the normal scientist is constrained by the disciplinary matrix from within which he attempts to solve the puzzles that are thrown up. His attitude is therefore largely uncritical and conciliatory if he is to make progress by solving puzzles within the disciplinary matrix. A failure to solve a puzzle is seen as a failure of the scientist rather than the inadequacy of the theory (1970b, p. 5). His task is to exploit the possibilities provided by the disciplinary matrix to the fullest extent, and as the esoteric details of his subject matter are further refined, so the disciplinary matrix becomes elaborated and extended in its application to the natural world. The attempt to subsume an increasingly large class of phenomena under the basic world view supplied by the evolving disciplinary matrix is a highly cumulative affair. Within the disciplinary matrix scientific knowledge grows by accretion.

Crisis and Revolution. The relatively trouble-free state of equilibrium in which the normal scientist works does not continue for ever. Occasional failures can be accommodated, a few unsolved anomalies set on one side so as not to hinder the progressive, puzzle solving activity within the disciplinary matrix. But there comes a time when the problems and difficulties encountered within the activities of normal science appear to undermine the framework provided by the disciplinary matrix in such a persistent and threatening manner that a crisis of confidence is generated. Kuhn offers as examples the crisis that led up to the Copernican revolution, and the problems associated with the aether and the earth's motion relative to it, towards the end of the nineteenth century. Such anomalies go against the expectations of the scientific community and persistent failure to solve such problems results in the 'pronounced professional insecurity' that sets in (1970a, pp. 67-8). He quotes W. Pauli's reaction to the confused state of Quantum Mechanics in the early 1920s. 'At the moment physics is again terribly confused. In any case it is too difficult for me, and I wish I had been a movie comedian or something of the sort and had never heard of physics.' That testimony supports Kuhn's observation if it is contrasted with Pauli's words less than five months later: 'Heisenberg's type of mechanics has again given me hope and joy in life. To be sure it does not supply the solution to the riddle, but I believe it is again possible to march forward' (1970a, p. 84). Kuhn offers a further psychological comment when he observes that 'Almost always the men who achieve these fundamental inventions of a new paradigm have either been very young or very new to the field whose paradigm they change.' The emphasis is not on the context of justification or on the validity of the theory, but on the psychology of the scientists involved in the 'break-down of the normal scientific community' (1970a, p. 90).

The increasing sense of insecurity in the scientific community is manifested by a breakdown of confidence in the once unifying disciplinary matrix. As individual members pursue their own 'extraordinary' research a proliferation of theories results.

> Confronted with anomaly or with crisis, scientists take a different attitude toward existing paradigms, and the nature of their research changes accordingly. The proliferation of competing articulations, the willingness to try anything, the expression of explicit discontent, the recourse to philosophy and to debate over fundamentals, all these are symptoms of a transition from normal to extraordinary research. (1970a, pp. 90-1)

If the crisis cannot be resolved, then ultimately a scientific revolution takes place. Considerable time might elapse between the first consciousness of breakdown and the emergence of a new disciplinary matrix and its supply of exemplars. But the revolution, when it comes, is sudden and catastrophic. Kuhn likens it to a 'gestalt switch' or a 'religious conversion' (1970a, p. 85). He offers no purely logical argument that demonstrates the superiority of one theory, or disciplinary matrix, over another which compels the rational scientist to make the switch. He simply refers to the psychological and sociological web of factors involved. Pursuing the political implications of the metaphor, he writes:

> It cannot be made logically or even probabilistically compelling for those who refuse to step into the circle . . . As in political revolutions, so in paradigm choice — there is no standard higher than the assent of the relevant community . . . This issue of paradigm choice can never be unequivocally settled by logic and experiment alone. (1970a, p. 94)

The reason for this is that the opposing camps in the divided scientific community do not share the same exemplars and thus attribute different meanings to the theoretical terms of the competing theories; they do not, in effect, speak the same language. Consequently the opposing camps, each arguing from within the values and standards of its own disciplinary matrix, are arguing at cross purposes. This is why the argument is one of persuasion rather than based on logic and experiment — although Kuhn insists that such arguments need not be irrational because there are rational means of persuasion.

If sufficient numbers switch their allegiance to the new theory and its accompanying disciplinary matrix then they will once again pursue normal science from within that matrix. Dissenters who remain outside the newly embraced orthodoxy are excluded from the new scientific community and they will eventually die out. 'The man who continues to resist after his whole profession has been converted has *ipso facto* ceased to be a scientist' (1970a, p. 159).

Progress in Science. Although the new disciplinary matrix possesses some of the symbolic generalisations of the one it has replaced, the theoretical terms take on new meanings in the new theory. For instance 'mass' and 'force' in relativity theory no longer have the meanings attached to them in the context of classical Newtonian Mechanics. Hence the scientific change which takes place in a revolution is not cumulative in the sense that the new theory is an improvement on the old. It is more in the nature of a reorientation, a new way of looking at the world which involves the rejection of the old science for the new. Although the scientific change within normal science is cumulative, the most significant, because drastic, change which results from revolution is non-cumulative. A conceptual change is involved: 'Though the world does not change with a change of paradigm, the scientist afterward works in a different world' (1970a, p. 121). Kuhn's fundamental point here is that one cannot escape from the framework; one sees the world through one's disciplinary matrix. And because the disciplinary matrix characterises the scientific community, so scientific knowledge is the collective opinion of such a community. He writes, 'I regard scientific knowledge as intrinsically the product of congeries of specialists' communities' (1970b, p. 253). Hence on Kuhn's account, scientific theories certainly undergo change, but do not progress relative to an end or goal. This is in direct contrast with Popper who, as we have observed, maintains that truth is the regulative principle by which theories are to be judged, and that as we approach the truth, 'so we get in touch with "reality"' (Popper, 1979, p. 360). Thus Kuhn writes, 'The normal scientific tradition that emerges from a scientific revolution is not only incompatible but often actually incommensurable with that which has gone before' (1970a, p. 103). In other words, there is no sense in which one disciplinary matrix with its attendant theory can be evaluated in terms of another.

Under pressure from his critics Kuhn insists that he is not adopting a relativist stance because there is a sense in which science can be said to progress. Although he denies that science, like evolution, is goal-directed, and that there is no sense in which scientific theories can be said to approach the truth and correspond more closely to reality, he maintains that progress can be monitored in terms of the greater problem solving ability of later theories over their predecessors. Thus Einstein is an improvement on Newton because his theory solves more problems than Newton's, although the two theories are, strictly speaking, incommensurable.

> Later scientific theories are better than earlier ones for solving puzzles in the often quite different environments to which they are applied. That is not a relativist's position, and it displays the sense in which I am a convinced believer in scientific progress. (1970a, p. 206)

Although Kuhn does not speak directly about the idea of truth in *The Structure of Scientific Revolutions*, his identification of progress in science with problem solving indicates his extreme scepticism with regard to truth. He implies that truth is a matter of consensus, or agreement, when he later writes, 'Members of a given scientific community will generally agree which consequences of a shared theory sustain the test of experiment and are therefore true' (1970c, p. 264). Under pressure from his critics he is a little more explicit. And contrasting his position with that of Popper he writes, '"truth" may, like "proof", be a term with only intra-theoretic applications' (1970c, p. 266). Thus he adopts an analytic view of truth, confining it to the puzzle solving activity of normal science. In other words, truth is confined within a particular framework or *Weltanschauung*; one cannot get outside one's framework and evaluate opposing theories in terms of truth. He reinforces this denial of any correspondence notion of truth when he writes in the 'Postscript', 'There is, I think, no theory-independent way to reconstruct phrases like "really there"; the notion of a match between the ontology of a theory and its "real" counterpart in nature now seems to me illusive in principle' (1970a, p. 206). Comparing his position with that of other philosophers of science he writes:

> They wish . . . to compare theories as representations of nature, as statements about 'what is really out there'. Granting that neither theory of a historical pair is true, they nonetheless seek a sense in which the later is a better approximation to the truth. I believe that

nothing of the sort can be found. (1970c, p. 265)

In the foregoing account of Kuhn's attempt to characterise the nature of scientific knowledge and change, it is evident that he favours a descriptive rather than a logical analysis of science. But like so many sociological accounts it is also prescriptive. Not only is Kuhn saying, 'This is what we regularly find in the history of science,' but also 'This is how, ideally, scientists ought to behave,' with the emphasis on normal science as opposed to extraordinary research and revolution. In other words Kuhn, through his attempt to produce a faithful historical picture, is in fact presenting a theory of science. It is not difficult to expose the reasons for this, for he must adopt theoretical assumptions before he can decide on what kind of activities to describe. He must be selective in his choice, and his selection is guided by his theory of what constitutes scientific activity.

The main ingredients of Kuhn's theoretical account have been outlined. The history of science displays episodes of normal science punctuated by occasional revolutionary upheaval. Normal science constitutes the bulk of scientific activity, and in Kuhn's estimation is more important. During normal science the scientific community, sharing the same disciplinary matrix or theoretical world view, is engaged in extending the applications of the theory and solving puzzles associated with it. Occasionally, when the puzzles become intractable and theoretical anomalies unresolvable, a crisis of confidence becomes apparent and alternative theories are considered. This leads to a scientific revolution in which there is a proliferation of various competing theories until one emerges as victor and is embraced by the new scientific community and new normal science emerges. John Watkins, in a critical review of Kuhn's thesis, nicely summarises Kuhn's view of the scientific community 'as an essentially closed society, intermittently shaken by collective nervous breakdowns followed by restored mental unison' (1970, p. 26).

This is the theory which Kuhn derives from his descriptive analysis of the scientific community. The concluding remarks of his 'Postscript' suggest 'the need for similar and, above all, for comparative study of the corresponding communities in other fields' (1970a, p. 209). Such a community relevant to our study would appear to be the religious community. Kuhn's characterisation of science has profound implications for religion which we shall subsequently examine. Indeed, John Watkins seized on the similarity between the scientific community and the religious community in Kuhn's thesis when he discussed Kuhn's preference for normal science.

> Kuhn sees the scientific community on the analogy of a religious
> community and sees science as the scientist's religion. If that is so,
> one can perhaps see why he elevates Normal Science above Extra-
> ordinary Science; for Extraordinary Science corresponds, on the reli-
> gious side, to a period of crisis and schism, confusion and despair, to
> a spiritual catastrophe. (Watkins, 1970, p. 33)

The assumption here, of course, is that the religious community is in
fact like the scientific community as characterised by Kuhn. A deeper
assumption concerns the uncritical use of the term *community* with
regard to such global categories as science and religion. We will return
to this matter in due course.

Kuhn's account of scientific knowledge has been the subject of
wide debate and severe criticism. First, despite his attempt at clarification, there are too many theses embedded in his notion of 'paradigm'
for it to retain much philosophical plausibility. Second, the distinction
between normal science and revolutionary science is questioned as a
matter of historical accuracy. Third, the incommensurability thesis
implies a non-empirical subjectivism that leads to irrationalism. Fourth,
if the scientist can only view the world through his disciplinary matrix,
or *Weltanschauung*, science is deprived of any sense of objectivity.
Finally, Kuhn's insistence that meanings are theory-dependent rests on
the mistaken assumption that theories are simply linguistic entities.

Kuhn has attempted to meet some of these criticisms, and sensitive
to the charge of irrationality he has modified his position, especially
with regard to the theory-dependence of observation, the theory-
dependence of meaning, and the comparison of competing theories. As
a result it appears that Kuhn has surrendered much that was distinctive
and original about his thesis, and has retreated to a position that re-
sembles the positivism he was concerned to reject.[5] It is perhaps as a
philosophical historian of science that Kuhn makes the most distinctive
and lasting contribution to our understanding of science, reminding us
that philosophers of science need to have a deep understanding of actual
scientific practice.

P.K. Feyerabend[6]

Our concern, however, is to pursue the implications of the *Weltanschauung* analysis of science which Kuhn develops in *The Structure of
Scientific Revolutions*, and which Feyerabend in particular pushes to

the extreme, without equivocation and with an uncompromising and unabashed enthusiasm. Although Feyerabend's starting point is the Popperian notion of criticism by falsification in science, his conclusions are radically opposed, not only to Popper, but to all attempts to provide a rational reconstruction of science. Popper attempted to explicate a methodology of science; to expose and recommend rules of procedure for the testing of scientific theories and so account for the growth of knowledge and its procession towards the ever elusive goal of truth. Feyerabend abandons all method in favour of what he calls 'epistemological anarchy'. He extends Kuhn's political analogy to describe science as in such a permanent state of revolution that his motto could well be described as 'Anarchy Rules, OK.' Let us trace briefly the route by which Feyerabend arrives at this radical and challenging conclusion.

The Interpretation of Theories. Feyerabend characterises what he calls *radical empiricism* as dependent on two fundamental assumptions which he identifies as the *consistency condition* and the *condition of meaning invariance*, both of which, he maintains, are untenable. According to the *consistency condition* new theories are consistent with their predecessors in a particular domain: that is, the new theory can be construed as containing the old theory, as for example it is contended that Einstein's theory contains Newton's as an approximation. According to the *condition of meaning invariance* the meanings of the common theoretical terms remain constant with respect to scientific progress. Feyerabend denies that these two conditions are met for the general, comprehensive theories of science such as 'the Aristotelian theory of motion, the impetus theory, Newton's celestial mechanics, Maxwell's electrodynamics, the theory of relativity, and the quantum theory' (1981, Vol. 1, p. 44). His argument concerning the consistency condition is that some of the consequences of the later theory are logically incompatible with the earlier theory. For example, Galileo's and Kepler's laws cannot be deduced from Newton's theory. Likewise with meaning invariance, Feyerabend claims that classical mechanics cannot be reduced to relativity theory because the term 'mass', for instance, has different and incompatible meanings in the two theories. 'The meaning of every term we use depends on the theoretical context in which it occurs. Words to not "mean" something in isolation; they obtain their meanings by being part of a theoretical system' (quoted by Shapere, 1981, p. 38). Because observation is also fully theory-dependent — 'observation statements have no "observational" core' (1981, Vol. 1, p. x) — there can be no neutral observation language by which theories can be tested objectively. No longer is science regulated by 'objective facts' because every single fact is dependent on some theory. Thus for Feyerabend, as well as for Kuhn, theories are *incommensurable* in the sense that 'the meanings of their main descriptive terms depend on mutually inconsistent principles' (1975, p. 277). Einstein can no more be compared with Newton than Darwin can be compared with Genesis.

Criticism, Proliferation, Truth and Realism. If there is no core of observational meaning which is common to all theories and which provides a basis for testing and comparing them, how can a theory be criticised? Feyerabend's answer is that the only way we can get any idea of what criticism would look like is to confront the theory with as many alternative incompatible theories as possible. The alternative theory will unearth facts which otherwise would have been suppressed, and so expose the theory to the fullest possible criticism. It is difficult, at this point, to understand Feyerabend's thesis of criticism by proliferation because if the theories in question are incommensurable, the facts of one theory can in no sense be relevant to another theory. In any case, Feyerabend insists that the choice between incommensurable theories is subjective. Ultimately we are are left with 'aesthetic judgements, judgements of taste, metaphysical prejudices, religious desires, in short, *what remains are our subjective wishes*' (1975, p. 285).

If our subjective wishes determine our theory choice, what becomes of truth and realism? Feyerabend abandons any notion of truth as correspondence to the facts, because there are no theory-independent facts. By the same token he rejects 'objective reality . . . [as] a metaphysical mistake' (1981, Vol. 1, p. xii). If we never have direct access to the world there is no point at which reality can in any sense enter into our theoretical constructions. Reality is constituted by our intellectual investigations, it is structured by our scientific theories rather than providing an independent target for them. 'We decide to regard those things as real which play an important role in the kind of life we prefer.' Realism is a ploy that the defenders of some system of thought adopt, it simply 'reflects the wish of certain groups to have their ideas accepted as the foundations of an entire civilization and even of life itself' (1981, Vol. 1, p. xiii). It seems, according to Feyerabend, that our subjective wishes determine everything, and the result is a complete relativism.

The links between Feyerabend and Kuhn are strong. Each views science as proceeding from within a *Weltanschauung* which determines how one views and interprets the world. Observations, and therefore facts, are totally theory-dependent. Meanings too are theory-dependent and will vary from one theory to another. Each theory makes its own world, and therefore the choice between a proliferation of incommensurable theories is arbitrary. Thus there is no hope of attempting to match the theory with reality, and the objectivity of science is a myth. Truth becomes a matter of agreement, and reality a chimera. Feyerabend, however, deplores the dogmatism in Kuhn's normal science and insists that science is, and ought to be, in a state of permanent revolution. Progress becomes change, and change becomes the replacement of one theory by another. Unlike Kuhn, who equivocates about irrationalism in science, Feyerabend has no qualms about his description of science as an irrational enterprise. 'One of the most important and influential institutions of our time is beyond the reach of reason as interpreted by most contemporary rationalists' (1981, Vol. 1, p. xiii).

Although much of Feyerabend's work is closely argued, often displaying a formidable mastery of his subject, he has not been very responsive to criticism levelled at his fundamental theses. And although his general conclusions are provocative and challenging, and provide a refreshing corrective to the contemporary worship of science, they are also somewhat bizarre, leading to extreme relativism and a subjective idealism that few find plausible. This is particularly apparent in his more recent publications *Against Method* (1975), and *Science in a Free Society* (1978), in which he probes the nature of science and attacks its status and prestige in Western society.

One of the fundamental reasons for this is exposed early in *Against Method*, and it concerns Feyerabend's use of language. He writes:

Incidentally, it should be pointed out that my frequent use of such words as 'progress', 'advance', 'improvement', etc., does not mean that I claim to possess special knowledge about what is good and what is bad in the sciences and that I want to impose this knowledge upon my readers. *Everyone can read the terms in his own way* and in accordance with the tradition to which he belongs. Thus for an empiricist, 'progress' will mean transition to a theory that provides direct empirical tests for most of its basic assumptions . . . For others, 'progress' may mean unification and harmony . . . *And my thesis is that anarchism helps to achieve progress in any one of the senses one cares to choose*. (1975, p. 27)

Here is an echo of Humpty Dumpty: 'When *I* use a word . . . it means what I choose it to mean — neither more nor less' (Carroll, 1871, Ch. 6). We may heed the Popperian warning against Essentialism, and we may grant the Wittgensteinian observation that links meaning with use, but language only works when people attempt to keep the rough and ready rules of customary usage. One who plays fast and loose with language cannot complain if his audience becomes confused. D.C. Stove (1982) points to the confusion generated by authors who constantly resort to quotation marks to neutralise familiar words such as 'success', 'truth', 'progress', 'objective' and 'rational'. It is difficult, for instance, to make much sense of Feyerabend's talk about observation when he writes, 'observation statements have no "observational" core' (1981, Vol. 1, p. x), because he is using the same word in one sentence in two different senses. Feyerabend and Humpty Dumpty are at liberty to make 'black' mean 'white', providing they are consistent in this usage. It is Feyerabend's inconsistency that leads to confusion: for instance, because we never quite know when a fact is a 'fact', we do not in fact know what he means! If Feyerabend adopts the role of Humpty Dumpty then he cannot complain if his readers, like Alice, find themselves in a wonderland.

Science and the Citizen. The theme of *Against Method* is well described by its subtitle: *Outline of an Anarchistic Theory of Knowledge.* Feyerabend's claim is that no successful rational reconstruction of science has ever been achieved. Strongly critical of the methodologies of Popper and Lakatos, he argues that the Western scientific tradition displays no particular methodological pattern, and that the most successful scientific enquiries have never proceeded according to the rational method at all. Dipping into history for evidence to support his claim, he details the arguments which Galileo employed to defend the Copernican revolution. He depicts Galileo not as a rationally minded scientist engaged in prediction, experiment and logical argument, but rather as a sort of political campaigner resorting to rhetoric, subterfuge and propaganda — traditional weapons of the anarchist. Such epistemological anarchy, maintains Feyerabend, is the only way to make progress in science. In the celebrated passage which enshrines this conclusion Feyerabend writes:

> It is clear, then, that the idea of a fixed method, or of a fixed theory of rationality, rests on too naive a view of man and his social surroundings. To those who look at the rich material provided by

history, and who are not intent on impoverishing it in order to please their lower instincts, their craving for intellectual security in the form of clarity, precision, 'objectivity', 'truth', it will become clear that there is only *one* principle that can be defended under *all* circumstances and in *all* stages of human development. It is the principle: *anything goes*. (1975, pp. 27-8)

Thus *Against Method* represents a frontal attack on the sacred bastions of rationality in science. Scientific research obeys no identifiable rules and does not operate according to methodological principles. The picture of science represented by the reconstructions of the logical empiricists, and the critical rationalism of Popper, is a myth; and what is more, maintains Feyerabend, it is a dangerous myth.

Having attempted to explode the myth that science is a rational enterprise possessed of a special method for understanding the world, Feyerabend next attempts to dethrone Western science as man's greatest success story, and to put it alongside other, equally legitimate, human ideologies and activities. Summarising this view he writes:

science is much closer to myth than a scientific philosophy is prepared to admit. It is one of the many forms of thought that have been developed by man, and not necessarily the best. It is conspicuous, noisy, and impudent, but it is inherently superior only for those who have already decided in favour of a certain ideology, or who have accepted it without ever having examined its advantages and its limits. And as the accepting and rejecting of ideologies should be left to the individual it follows that the separation of state and *church* must be complemented by the separation of state and *science*, that most recent, most aggressive, and most dogmatic religious institution. Such a separation may be our only chance to achieve a humanity we are capable of, but have never fully realized. (1975, p. 295)

Thus Western science is just one tradition amongst many, with no special claim to authority. This is a theme which Feyerabend follows up in *Science in a Free Society* (1978), in which he attempts to undermine the status and prestige and authority of Western science. Here he examines the political and economic factors that combine to prop up the great Establishment of science. As a disciple of Mill, and as an admirer of his 'immortal essay *On Liberty*', he supports 'the attempt to increase liberty, to lead a full and rewarding life, and . . . therefore, the rejection of all universal standards and of all rigid traditions' such as a large part of contemporary science (1975, p. 20).

Therefore Feyerabend not only argues for a proliferation of theories within science, but for a proliferation of *Weltanschauungen*, or world views, to counter the oppressive influence of the *Weltanschauung* characterised by rationalist epistemology. Such pluralism is necessary if an individual is to have the best chance of genuine freedom of choice between competing ideologies and traditions. The citizen confronted with alternative ideologies stands a better chance of making a rational choice than the citizen instructed in a single, special ideology. Feyerabend argues that our education system should be more liberal and flexible, open to alternative systems of thought. He does not see why, if there is a choice concerning religion, there should not be a choice on the curriculum between magic and science. He does not deny the existence of genuine freedom within science in Western democracies; his argument is against the dominance of the rationalist tradition at the expense of other, non-rationalist, traditions. Rather than subsume all human affairs under some all-embracing unified system of thought and practice, he advocates a plurality; a policy of 'live and let live', of diversity and non-conformity in the place of unity and uniformity. Therefore just as the theories of science are incommensurable, so should the distinctiveness of competing ideologies and traditions be maintained. Feyerabend is therefore 'against the thinning out of subjects so that they become more and more similar to each other' and suggests, for instance, that

> whoever does not like present-day Catholicism should leave it and become a Protestant, or an Atheist, instead of ruining it by such inane changes as mass in the vernacular. This is true of physics, just as it is true of religion or prostitution. (1975, p. 308)

Choice is what matters; choice between clear alternatives. And the greater the choice the greater the freedom for the individual to embrace a well-considered view.

The Paradox of Feyerabend. There is something paradoxical about Feyerabend's overall approach. First, although he is intent on debunking science and the rationalist tradition, his method of argument and presentation, particuarly in his more technical philosophical papers, is deeply entrenched in the rationalist tradition he disdains. Second, in recommending a plurality of commitments and values there is a sense in which he is articulating a thesis which is itself an all-embracing ideology. Third, his slogan 'anything goes' can, in a perverse manner, itself be construed as a methodology, and a conservative one at that! We have already emphasised Popper's warnings regarding the abandonment of the critical-rational approach in science, and its disastrous philosophical and even political consequences. If Feyerabend's view is adopted, and everyone is encouraged to pursue his own inclinations, with a complete disregard for rational criticism and appraisal, then those who have power will tend to cling to it. John Krige points out that Feyerabend's epistemological anarchism, far from being liberating, is highly conservative with regard to change. He comments, 'anything goes . . . means that, in practice, *everything stays*' (Krige, 1980, p. 142). This is hardly what Feyerabend wants to advocate.

But this only deepens the paradox, for although Popper is Feyerabend's *bête noire* and the frequent subject of flippant abuse, there remains a strong undercurrent of Popperianism in his views. He has inherited from Popper (and Mill) the strong advocacy of criticism which, he insists, must be given free rein among a proliferation of theories and traditions. Further, this argument for intellectual democracy is similar to Popper's argument for political democracy with its emphasis on the freedom of individuals to change the existing system. In fact, this might be the key to understanding Feyerabend's insistence on proliferation in general, and his rejection of methods and rules which cramp and restrict. Proliferation furthers freedom – but freedom for what? Feyerabend's reply is freedom to criticise; but not in order to approach the truth or to obtain a deeper understanding of the world. Maybe it is just that freedom is more fun; that variety is the spice of life! Proliferation, it seems, is an end in itself. Yet, as we have pointed out, unrestricted proliferation, and freedom divorced from all constraint, leads inevitably to the very authoritarianism it seeks to avoid.

However, what Feyerabend like Popper, has done, is to demonstrate that epistemology matters, and that it has far-reaching implications for our lives. By attempting to expose the nature of scientific thinking, albeit as one myth among many, Feyerabend has opened up some of the wider issues of politics and religion which deeply affect the way we live our lives. Feyerabend alludes to these deeper concerns in his recent introduction to some of his collected philosophical papers when he suggests that 'the abstract ideas that the intellectuals have tried to put over on us ever since the so-called rise of rationalism in the West' might be 'of only secondary importance when compared with the conditions of the soul as described in religious beliefs' (1981, p. xi).

The Fundamental Flaw of the Weltanschauungen *Analyses*

Before we turn to the implications of the *Weltanschauungen* analyses of science for religion, we must direct a little more critical scrutiny at the underlying features of the epistemology that yields such analyses.

Three related epistemological theses are embedded in the *Weltanschauungen* analyses of science, concerning *observation, meaning* and *fact*. Each of these theses has a strong and a weak version. The weak version consists in the recognition that our *observations* are influenced by our theories, prejudices and presuppositions; that the *meanings* of the terms of a theory are not totally independent of the theory in which they are contained: that therefore what counts as a *fact* is partially determined by theoretical considerations. Such a view derives (in modern philosophy) from Kant, who concluded that there is a relationship between the structure of the human mind and the world. It was the admission of these three weak theses that undermined the Received View and which contributed to the Popperian account of science. We have discussed Popper's attempt to provide an analysis of science which attempts to accommodate these three weak theses within a flexible methodology without renouncing truth, objectivity and realism. The three weak theses alone do not imply relativism unless they are coupled with the separate thesis that different theories carry with them different *Weltanschauungen*.

However, it is the strong version of these three theses which allows of no compromise, and which leads to a complete relativism and an extreme subjective idealism, which is implicit in the *Weltanschauung* analysis of science offered by Kuhn and Feyerabend. Accordingly, *our theoretical presuppositions completely determine how we view the world*: different *Weltanschauungen* yield different observations. When Feyerabend says that observation has no 'observational' core he can only be asserting that what people have traditionally meant by observation is a complete myth. *Meanings also are totally theory-dependent*: any change in theory alters the meanings of all the terms of the theory. Therefore, *each theory produces its own facts*. Because the meanings have changed, competing theories are not just inconsistent with each other because there is no way a comparison could be made: they are incommensurable. Our subjective wishes are the deciding factor in the adoption of one theory in place of another.

In the *Weltanschauung* approach to science, observation becomes 'observation', meaning becomes 'meaning', and facts become 'facts', where the inverted commas signify not just that we qualify and clarify these terms, but that they have become totally obsolete. A profound and enlightening thesis of Kant has become an extreme and uncompromising dogma. And it is only when we formulate the thesis in this extreme form that we perceive that it is self-defeating and absurd. It is self-defeating because, if true, there would be no way of formulating the thesis in a significant manner. That is why, in spite of some of his profound analysis and important insights, it is difficult to understand what Feyerabend is propounding. When practically all the key words in the philosophy of science are put in quotation marks we are at a loss to know what is being said. When, for instance, 'progress' becomes 'change' or 'replacement', why bother with the word 'progress'? And why bother with 'truth' and 'realism' if truth and realism are denied?

Nevertheless, various versions of these three theses have been subject to damaging attack. The first thesis, regarding observation, embraces a subjective view of knowledge which is incompatible with the alleged objectivity of scientific observation. If proponents of different theories cannot observe the same things in an effort to decide between the theories, then there can be no inter-subjective testing. A paradoxical view of science results in which each viewpoint creates its own reality. Various arguments have been formulated in an attempt to refute such a subjective idealism, and although it is debatable whether or not a conclusive refutation is possible, such a subjective idealism has little explanatory power to commend it as a plausible thesis.

Likewise with the third thesis concerning the relationship between facts and theories. If what counts as a fact is determined by the *Weltanschauung* associated with a theory, then it follows that a theory cannot be objectively tested. A subjective view of science results which denies the working assumption that science deals with a common world. The fact that such a view is unacceptable to, and inconsistent with, working science does not disprove it. But if we accept it, we must, with Feyerabend, conclude that we are imprisoned by our framework or *Weltanschauung*, and surrender any claim to rationality. Each *Weltanschauung* sets its own standards, and there is no way in which they can be compared or evaluated.

However, the central issue is that which concerns meaning change and incommensurability. Following Suppe, we will summarise the consequences of this thesis (Suppe, 1977, pp. 199-208). First, Feyerabend's *theory*, like Kuhn's *paradigm*, is such a wide, all-embracing affair, including ordinary beliefs, myths, religious beliefs, in fact any sufficiently general point of view, that it is not at all clear what would constitute a change in theory. Does a slight change of belief or the revaluation of a physical constant constitute a change of theory and therefore a change of meaning? The question of meaning change leads us into the intricacies of the philosophy of language and exposes the difficulties of establishing anything like a criterion by which we could identify a change of meaning. Second, if the meaning of a term is determined by the theory in which it occurs, then the same term in different theories will have different and incompatible meanings. Consequently the theories can never contradict one another. Third, if different theories can neither agree nor disagree, in what sense can they be viewed as alternatives between which a choice is to be made? Fourth, if meaning is theory-bound, then truth is analytic, and it follows that science loses its empirical character. Fifth, all theory testing will be circular because the only observations relevant to the theory will be those which are consistent with the theory.

Although the *Weltanschauungen* analyses arose as a reaction, and an alternative, to the Received View, both views attempt to discover the nature of scientific theories through an examination of their linguistic formulations, and even at times imply that the theory *is* its linguistic formulation. This is not a far remove from Wittgenstein's attempt in the *Tractatus* to establish that the logical structure of the world is mirrored by a logically perfect language. It is because Kuhn and Feyerabend construe theories as linguistic entities that they are misled by problems of meaning. However, once it is realised that theories are extra-linguistic entities, and not just sets of propositions, the problems concerning the analysis of meaning become irrelevant to the question of theory change. This is easily demonstrated by translating a theory from one language into another, for by so doing we produce a different set of propositions without altering the theory. Likewise there are many theories in physics which can be formulated in different ways for different purposes; again, the same theory is expressed in different sets of propositions. Thus, there is a sense in which language is plastic to the theory.

In summary, then, the reason why paradigm or theory differences are considered incommensurable is because the sets of *meanings* contained within the theories are incommensurable. There is no way in which they can be directly compared or critically evaluated. The assumption here is that any shared tradition or point of view involves identity of meanings. The only alternative to such identity is complete difference, which leads to total logical incompatibility or incommensurability. There is no middle ground between these two exclusive extremes, and therefore the rejection of the positivist condition of meaning invariance lands one in the relativist camp. Shapere observes that this relativism, and the doctrines which eventuate it, is not the result of an investigation of actual science and its history; rather, it is the purely logical consequence of a narrow preconception of what 'meaning' is. Nor should anyone be surprised that the root of the trouble, although not easy to discern until after a long analysis, should turn out to be such a simple point, for philosophical difficulties are often of just this sort. (1981, p. 55)

Whether or not the mistake on which this relativism is founded is a simple one, the preoccupation of philosophers with meaning has certainly had a long and influential history with widespread repercussions. Popper recognised the futility of a philosophy based on meaning in his long-standing battle with Essentialism, and he untiringly directs our attention to the goal of truth or greater explanatory power, rather than to the niceties of meaning. Russell, too, had little time for the post-Wittgensteinian preoccupation with meaning, and it is significant that he did practically all his important philosophical work when he 'thought of language as transparent — that is to say, as a medium which could be employed without paying attention to it'. He continues, 'The essential thing about language is that it has meaning — i.e. that it is related to something other than itself, which is, in general, non-linguistic' (Russell, 1975, p. 11). Shapere, in his critical review of the relativism of Kuhn and Feyerabend, arrives at the same conclusion: that we should not be misled by talking about the meanings of words in our attempt to understand the structure and function of scientific theories. For the purpose of establishing the central features of scientific theories, and of comparing different theories, there is no need to talk about meanings. This is not to say that we should not seek clarity of expression and, if necessary, at times, specify what would count as a change of meaning, but

> *if* our purpose is to understand the workings of scientific concepts and theories, and the relations between different scientific concepts and theories — if, for example, our aim is to understand such terms as 'space', 'time', and 'mass' (or their symbolic correlates) in classical and relativistic mechanics, and the relations between those terms as used in those different theories — then there is no *need* to introduce reference to meanings. And in view of the fact that that term *has* proved such an obstruction to the fulfilment of this purpose, the wisest course seems to be to avoid it altogether as a fundamental tool for dealing with this sort of problem. (Shapere, 1981, p. 57)

Shapere concludes that the thesis of the theory-dependence of meanings (held by relativists) and its opposite, the condition of meaning invariance (held by positivists), derive from the same mistake (or excess). There is, of course, some truth in each thesis, and that is why the weak versions of the theses considered are important; but this must not lead us into the error of concluding that there can be *no* resemblances between two sets of terms. Clearly, there are differences of meaning between Einsteinian and Newtonian terms, but there remains considerable overlap which both indicates the continuity and allows us to compare theories rationally.[7] The error consists in supposing that unless there is complete identity there must be absolute difference. And that error in turn derives from a mistaken philosophy of the functions of language which we discussed in the last chapter.

There remains the overall question concerning the viability of the *Weltanschauung* thesis and its relationship with scientific theorising. Are the members of a scientific community, or tradition, bound by the joint possession of a common *Weltanschauung* or conceptual point of view? Are they all confined within a framework outside which rational and fruitful discussion is impossible? On their own admissions Kuhn's paradigms and Feyerabend's theories are exceedingly complex, all-embracing entities, beyond adequate description, including background, training, experience, knowledge, beliefs and points of view. As such, the *Weltanschauung* is such a metaphysically and epistemologically inflated affair that it is difficult to see how it can be the *common* possession of a group of scientists, or for that matter of any other group. The features mentioned as constituents of the *Weltanschauung* are simply the features that mark off one individual from another. Even two individuals working in close co-operation will not share precisely the same *Weltanschauung*, or wear identical conceptual spectacles; or to state a truism, no two individuals are identical or have identical outlooks. Thus even to postulate the possession of individual *Weltanschauungen* is superfluous. We can, if needs be, draw attention to specific characteristics of a working scientist, such as his use of language, or his beliefs, and indicate their relevance to his contribution to theory formation, but to refer to every item of his scientific make-up as a basic unit in the philosophy of science is both unnecessary and unhelpful.

What is characteristic of scientists in a given tradition is that there exist similarities and differences concerning theoretical and methodological matters. The similarities do not have to be identities for the work to proceed; and the differences do not need to be divisive to the extent that communications break down. This is a matter of plain historical and sociological observation. Thus the assertion that a scientific community is contained by the joint possession of a *Weltanschuung*, which can only be replaced by some ideological revolution, is not derived from historical and sociological study of science, but is based on a logical error.

In summary, the epistemological theses that underlie the *Weltanschauungen* analyses of science imply a subjective idealism which is incompatible with the declared objectivity of science. The doctrines concerning the theory-dependence of observation, meaning and incommensurability deny the possibility of rational theory assessment. The view is fundamentally defective since it makes discovery of how the world really is irrelevant to scientific knowledge, reducing scientific knowledge to the collective prejudices of members of scientific communities.

In his 'Afterword' to *The Structure of Scientific Theories* (1977), Suppe concludes that 'the *Weltanschauungen* analyses are not widely viewed as serious contenders for a viable philosophy of science ... the *Weltanschauungen* views, in a word, today are *passé*' (pp. 633-4).

Although the *Weltanschauungen* analyses (particularly the historical investigations of Kuhn and Feyerabend) have enlightened our understanding of the context of discovery, by pointing to the relationship between theory and scientific community, they do not contribute much to our understanding of the *structure* and *function* of scientific theories, their explanatory power and their relationship with the world — and this is the central task of the philosophy of science.

Ludwig Wittgenstein

We have observed that the root of the relativism of Kuhn and Feyerabend, and of the *Weltanschauungen* views in general, is a conceptual one, arising from a doctrine of meaning. Meanings are bound by the general theories which in turn bind communities within a common framework or world view. This preoccupation with meaning has had a long history which can be traced back at least as far as Plato, but in more recent times the overwhelming influence has been that of Ludwig Wittgenstein. It is commonly asserted that Wittgenstein is unique amongst philosophers because he produced two major philosophies in his lifetime, contrasted in his two major works, the *Tractatus* (1922) and the *Philosophical Investigations* (1953). In the first he attempted to find a precise logical match between language and the world and to define the limits of language; in the second he rejected this as a mistaken and misguided effort, and directed his attention to the way in which language is used. His abandonment of the referential theory of meaning of the *Tractatus* for the theory of 'meaning as use' represents a movement away from the analysis of propositions to the examination of the *use* of language as part of a whole complex form of life. And this shift of philosophical interest is precisely reflected by the movement from the analysis of scientific theories to the sociological examination of the way theories are developed by scientists. But for all their contrasting features and contrary conclusions, both are concerned with language and meaning, and in particular with the drawing of linguistic boundaries. This persistent concern with meaning is the thread that links all Wittgenstein's thought, and one can read the *Philosophical Investigations* and subsequent posthumously published works as a sequel to, and a development of, his earlier thought in the *Tractatus*. Wittgenstein, in fact, wanted to include the text of the *Tractatus* alongside the *Philosophical Investigations* because of their close affinity (see D. Pears in Magee, 1971, p. 45). The same thread links the Received View with the *Weltanschauung* view of science. And just as the *Tractatus* proved to be a profound influence in the development of Logical Positivism, so Wittgenstein's later ideas were a predominant influence in the work of a whole generation of philosophers which resulted in the sort of relativism we have discussed. So there is a sense in which relativism is a direct descendant of Logical Positivism, and one of the founding fathers of both is Ludwig Wittgenstein.[8]

There is little evidence for this conclusion in the writings of the proponents of the *Weltanschauungen* analyses of science, who present their theses as a clear refutation and denial of positivism. Kuhn and Feyerabend, for instance, lay great stress on history, and claim that their ideas derive from examples of actual scientific practice. And yet the predominant philosophical influence at the time of Kuhn's researches for *The Structure of Scientific Revolutions* was provided by the legacy of Wittgenstein. For it is in his post-*Tractatus* work that Wittgenstein develops the notions of the *language-game* and the *form of life*, which are germane to Kuhn's notion of *paradigm* and his theory of 'meaning as use' which both Kuhn and Feyerabend exploit to the full in their theses of meaning change. As his posthumously published notebooks and letters show, Wittgenstein was developing these ideas in the 1930s and refining them to the time of his death in 1951. As a result, a whole generation of philosophers latched on to his ideas and attempted to apply them in such areas as science, politics, ethics, sociology, education and religion. The development and application of Wittgenstein's ideas, initially by his students at Cambridge, and subsequently by later philosophers, gave rise to a powerful and influential school of thought in which philosophical problems were dissolved by a study of their grammar.[9] The attempt to characterise science in terms of a *Weltanschauung* analysis is intellectually rooted in Wittgenstein's philosophy, and that is doubtless one of the main reasons why the approach was received with such acclaim and attention. Commentaries on Wittgenstein's later philosophy are legion, and what follows is but a brief pointer to those features of his philosophy which are of obvious relevance to our thesis.

Language-games and Forms of Life. In the *Philosophical Investigations* Wittgenstein rejects the thesis of the *Tractatus* that there is only one single language of science which corresponds with reality, and in its place introduces the view that language consists of a multitude of different, and often interacting, *language-games*, each with its own grammar or rules of use. By 'grammar' here, Wittgenstein does not mean the surface grammar which characterises the structure of meaningful sentences, but the depth grammar, or the underlying logic of language. Confusion arises particularly from mixing the grammars of different language-games: that is, from applying the rules of usage in one language-game to another. Language trespasses its limits, or 'goes on holiday', when particular expressions are used outside their proper domain or range of application. In other words, when the rules are broken philosophical puzzles arise, and these puzzles are only dissolved when order is restored. In his review of the multiplicity of language-games Wittgenstein cites, amongst many examples, 'forming and testing a hypothesis' and 'praying' (PI 23).[10] Thus confusion is generated when beliefs about God are treated as scientific hypotheses. Gilbert Ryle pursued the same line of thought in his *Concept of Mind* when he referred to 'category mistakes' that generate confusion by transferring a concept from one category to another. And elsewhere Ryle makes the same point when he writes: 'If the seeming feuds between science and theology . . . are to be dissolved at all, their dissolution can come not from making polite compromises but only from drawing uncompromising contrasts between their businesses' (1954, p. 81).

Central to Wittgenstein's later philosophy is the idea of a *form of life*; a term which embraces the idea of language-games and rules of usage with a common way of life. Thus he describes the language-game as 'the whole, consisting of language and the actions into which it is woven' (PI 17). 'To imagine a language', writes Wittgenstein, 'is to imagine a *form of life*' (PI 19). And again: 'the term "language-*game*" is meant to bring into prominence the fact that the speaking of a language is part of an activity, or a form of life' (PI 23). His fundamental point here is that the use of language is part of a wider human activity or form of life. Language is not intelligible if it is set apart from the conceptual scheme and cultural activity of a common group. That is why people who do not share the same forms of life or cultural traditions as ourselves can be 'a complete enigma to us'. This is the significance of Wittgenstein's aphorism, 'If a lion could talk, we could not understand him' (PI 223). Talking lions, like talking parrots, would only be echoing human speech, and not intelligently or significantly participating in our form of life. If we are to understand each other, then we must share something of each other's conceptual schemes and ways of life. The language-games that constitute science and religion are mumbo jumbo to those uninitiated into their respective forms of life. Thus a scientific theory or a credal statement is only intelligible within its respective form of life; that is, within the context of how its proponents 'think and live' (PI 325). And each form of life is self-authenticating; it provides its own justification. We can no more distance ourselves from a form of life of which we are a part than we can climb out of the universe to examine it.

Language and Reality. In contrast to the *Tractatus* view that language pictures the world or corresponds to reality, Wittgenstein develops the idea in his later writings that reality is in some measure constituted by language. There is a sense in which we make the world with our language. 'Grammar tells us what kind of object anything is. (Theology as grammar)' (PI 373). And 'grammar' here, as we have indicated, refers to the underlying logic of language. People with different 'grammars' see the world differently. Waismann suggested that the German who says, 'The sky blues' sees something different from the Englishman who, pointing to the same sky, says, 'The sky is blue.' One is seeing colour as an active agency, and the other as the passive property of a surface. There is a subtle interplay between language and the world, and it is impossible to distinguish between *our* construction of reality by the way in which we use language, and the world which gives rise to our linguistic constructions.

This is one of the central ideas we have traced in the work of Kuhn and Feyerabend who insist that the scientist, rather than being the detached investigator of the world of facts, helps to shape the world he investigates. Our conceptual perspective is intimately linked with the way our language conceptually shapes our understanding of the world. Thus different communities, operating with different paradigms or theories, will see the world differently. And Feyerabend, as we have noted, uncompromisingly rejects objective reality as a metaphysical mistake. Reality, he insists, is structured by our theories.

Winch, following Wittgenstein, appears to be more uncompromising in his refusal to separate 'reality' from language, so that language actually appears to determine what is real. He writes:

> Reality is not what gives language sense. What is real and what is unreal shows itself *in* the sense that language has. Further, both the distinction between the real and the unreal and the concept of agreement with reality themselves belong to our language. (1967, p. 13)

Wittgenstein's discussion of the relationship between language and reality is both more subtle and more inconclusive than some of his followers have inferred from his writings, and he avoids the subjective idealism which is implicit in the work of Kuhn and Feyerabend. But Popper's criticism, that Wittgenstein's preoccupation with the medium of language prevents him from using language to investigate problems thrown up by the world, is very much to the point. Preoccupation with meaning inhibits the use of language as a critical tool for the investigation of the world with the aim of discovering the truth. Donald Hudson, who has attempted to explore and explicate the implications of Wittgenstein's philosophy for religion, will have none of this; philosophy is not about understanding the world, but about understanding language. He writes: 'What else could understanding be conceivably concerned with, if not with words? And understanding is the goal of philosophy' (Hudson, 1975, p. 48).[11]

Although Wittgenstein reckoned that his concern was to offer descriptive studies of language-games and to show how meaning derives from use within the socially fixed parameters of a language-game, he has in fact presented a philosophical theory of meaning. The philosophical thesis may not be as explicit as that of the *Tractatus* but it is based on the enormous assumption that philosophical problems can only be tackled through an examination of meaning; that is through an examination of the *use* of language. It was this philosophical thesis that inspired the philosophy of conceptual analysis which dominated the philosophical scene for a generation or more, and which ultimately led to the relativism implicit in the *Weltanschauungen* analyses of science. It is from Wittgenstein's insistence that his work is descriptive that it follows that criticism is prohibited. Language-games may not judge one another. 'Philosophy may in no way interfere with the actual use of language; it can in the end only describe it . . . it leaves everything as it is' (PI 124). There is a sense of ultimacy beyond which we cannot go. The language-game, the form of life, contains within it its own justification, and all one can say is 'This language-game is played' (PI 654).

Wittgenstein's *form of life*, with its embedded notion of language-game, is as central to his philosophy as Kuhn's *paradigm* is to his. The two terms are in many ways almost interchangeable. Both refer to a 'community's shared belief' (Kuhn, 1970a, p. 43), and both are equally vague and all-embracing. The confusion which surrounds Kuhn's use of *paradigm* is matched by that which derives from Wittgenstein's *form of life*. A common feature is their self-contained nature. They are intelligible only to those truly initiated into the culture, and cannot be understood from the outside. As Roger Trigg comments, 'To dub something a "form of life" is in effect to protect it from criticism' (Trigg, 1973, p. 66). Winch effectively erects such protective barriers around religion and science respectively when he insists that each is a different mode of social life (Trigg, 1973, p. 60). Each provides its own criteria of intelligibility and cannot be judged from outside.

The similarity with Kuhn and Feyerabend is striking. The paradigm, or the theory, like the form of life, tends to be self-contained and insulated against criticism. Kuhn's study of the scientific enterprise leads him to the discovery of normal science, constituted by its paradigm. 'This is what we find,' maintains Kuhn. Likewise Wittgenstein refers to the form of life as 'the given' which 'has to be accepted' (PI 226). So for Wittgenstein and for Kuhn, Western science is a form of life; spectators, as critics, are not allowed. One must play the appropriate language-game according to the rules, or operate within the paradigm, the form of life, in order to understand it, in order to have concourse with the members that constitute its society. There are no external, objective, rational standards such as truth or reality, by which the community's beliefs and practices may be judged. The form of life, like the paradigm, authenticates itself. It is its own validation.

Feyerabend, in a postscript to an article on Wittgenstein's *Philosophical Investigations*, has commented on the connection between Wittgenstein's language-game and form of life and Kuhn's paradigm. 'Neither can be understood on the basis of simple and abstract descriptions, 'we must move away from traditional concepts of clarity and precision in order to arrive at a more realistic picture of scientific change; just like a language-game a paradigm is not a well defined entity but a word for a practice whose elements become known only to those who participate in it. Thus no theoretical account of a science is possible. 'All sciences are social sciences — even mathematics — there are no actual sciences.' Feyerabend enlists Wittgenstein's support to sustain his contention that we cannot understand science by studying its rationality, that is by identifying its characteristic laws, standards and rules. We can only understand science by participation, by practising it, that is, by engaging in its form of life (Feyerabend, 1981, Vol. 2. pp. 129-30).

Paradigms, Forms of Life and Communities

We have observed that central to Kuhn's account of science is the scientific community, bound in social and intellectual cohesion by its paradigm; and we have noted his recommendation for 'the need for a comparative study of the corresponding communities in other fields' (Kuhn, 1970a, p. 209). And we have suggested that one of the fatal weaknesses of this *Weltanschauung* view lies in the assumption that there are in fact such identifiable communities, or in Wittgensteinian terms, forms of life.

There are doubtless small communities of scientists that are closely bound and involved in a common enterprise, using a specialised vocabulary, and engaged in pursuing a common goal. One could cite as examples communities of scientists gathered around the world's great particle accelerators, as at CERN. Likewise there are small religious communities, bound, as in the religious orders, by their common vows and shared way of life. But to lump together all 'scientists' in some great 'scientific community', or all 'religious' people in some common, identifiable 'form of life' is a nonsense. Sociologically there is no evidence that such well-defined and self-contained communities exist; and intellectually the idea is absurd. Saifullah Kahn, commenting on multicultural relationships in Britain, makes much the same point when she suggests that Muslims are more accurately described as a category than a community; that is a group sharing certain characteristics in common rather than one participating in a single, common structure (or closed form of life). She argues that it is the host society which is responsible for viewing them as a community, and thus as a threat to the indigenous population (1976, pp. 227-8).

In all the critical reviews of Kuhn's seminal book, *The Structure of Scientific Revolutions*, no one appears to have seized upon this gigantic

shit

fallacy. Kuhn has swallowed the bait that has hooked so many of our contemporaries dominated by the so-called twentieth-century scientific world view, that somehow all scientists are alike; that they all belong to some modern high priestly cult, the great community of scientists. Nothing could be further from the truth, and ironically it is based on the popular error which Kuhn himself is at pains to reject, that there is some such thing as the 'scientific method' which binds all scientists and which, once acquired, will lead to all knowledge and all truth.

The fact is that there is a whole range of so-called sciences: from the social and psychological sciences, to the life sciences, to the physical sciences, to the highly abstract and esoteric mathematics involved in the enquiries into the fundamental structure of matter. And there is a whole range of so-called scientists who adopt such a wide variety of methods and outlooks that it would be tedious and pointless to attempt to catalogue them all.

Likewise with religion. Not only are there the great monotheistic world religions of Judaism, Christianity and Islam, there are also the ancient religions of India and China, and a thousand and one modern American sects. And many of these religions exist in widely different social systems: Christianity in Russia and the United States and in South Africa; Islam in Iran and Pakistan; and so on. There is no way of identifying all religious people as belonging to *a* community of believers, or to a characteristic form of life. Different religions such as Christianity and Islam can be studied for specific purposes such as their doctrinal formulations; and different expressions of the Christian or Muslim way of life can be investigated. What is more, within each identifiable religion there exist considerable differences of interpretation, and fundamental disagreements on important matters such as capital punishment, birth control and abortion. It is doubtful if a congregation of one hundred people reciting the Nicene Creed will all share a common conceptual view of their beliefs.

Further, as Trigg points out,

It is unimpressive to be told that a certain . . . disagreement can only be explained by the difference in forms of life from which those disagreements come, and then to find out that the disagreement itself is the only criterion for identifying the two forms of life. (Trigg, 1973, p. 66)

Here is circularity indeed!

However, if we grant that within Western science are a number of ill-defined, but related, forms of life, then by the same token religion too can be construed as a system of interacting forms of life bound by a common thread. But according to Kuhn they are impervious to criticism from outside their respective paradigms or, as Wittgenstein would have it, from outside their respective forms of life. One can no more externally judge the conceptual scheme that constitutes religion than one can criticise the whole conceptual scheme that constitutes Western science. What Gellner said about Wittgenstein's later philosophy with respect to religion is equally applicable to science. 'By destroying philosophy, Wittgenstein made room for faith . . . religious believers can find in Wittgensteinianism not merely a device for ruling out philosophic criticism, they can find in it a positive validation of their belief' (quoted by Bartley, 1974, p. 125). Commenting on his later philosophy Wittgenstein remarked, 'Its advantage is that if you believe, say Spinoza or Kant, this interferes with what you believe in religion; but if you believe me, nothing of the sort' (quoted by Hudson, 1968, p. 67).

It is the rejection of criticism by Wittgenstein, by attempting to limit philosophy to description, that contrasts with the traditional role of philosophy. It contrasts sharply with Popper's contention that the philosopher's main function, and indeed the criterion of rationality, is that of criticism. One is a *laissez faire* philosophy which, on the face of it, is simply an uncritical recognition and acceptance of a variety of widely differing and non-interacting forms of life; whilst the other adopts an active, problem solving, truth-seeking attitude to the world which is the ever fascinating object of man's inquisitiveness and enquiry.

One of the problems regarding Wittgenstein's philosophy concerns its interpretation. Wittgenstein was never happy with Russell's introduction to the *Tractatus*, nor with the use the Logical Positivists made of his thesis. Doubtless the reason for this has much to do with Wittgenstein's own dissatisfaction with his work. And this is true of his late philosophy, which, as we have observed, arose directly out of his disaffection with his earlier ideas. For such reasons Wittgenstein was loath to publish his later ideas and he admits in the preface to the *Philosophical Investigations* that it was the best offering he could make in the circumstances. Since his death a whole rash of books bearing his name have appeared, containing practically every one of his jottings and comments recollected by his students.

There is no doubt that this later work of Wittgenstein gave rise to a new way of doing philosophy which was itself a combination of fairly widely differing interpretations of the master's thoughts, and we have suggested that the new direction given to the philosophy of science by the *Weltanschauungen* analyses was an influential and important example. However, there have always remained those who were dubious about the merit and value of Wittgenstein's later work, and some who have viewed its influence with considerable misgivings. Russell confessed that he viewed Wittgenstein's later philosophy as trivial and somewhat pointless, and Popper maintains that it is vastly overrated and largely a waste of time and effort and really rather boring. Peter Munz maintained that it had a bad pedagogic effect and that consequently it misled a whole generation of students of philosophy.[12]

Wittgenstein and Religious Belief

However that may be, the influence of Wittgenstein is undisputed, not least in the philosophy of religion, and it is our contention that this 'fideism', as it has been labelled, is almost a mirror image of the philosophy of science contained in the *Weltanschauungen* analyses we have discussed. There is no doubt that deeply embedded in all the philosophical perplexity that characterised Wittgenstein's life and work was a deep respect for religion. The last remark of the *Tractatus*, 'Whereof one cannot speak, thereof one must be silent' (7), far from being a neat, conclusive dismissal of religion as nonsense, is more accurately construed as a vindication of the mystical which is at the heart of all religion.[13] The significant change of attitude with regard to religion, on Wittgenstein's part, is evidenced not so much by a change of heart towards religion but by his realisation that religion *can* be a significant subject for discourse. Thus his 'Lectures on Religious Belief' (1966), given a few years after the completion of the *Tractatus*, are a vindication of the significance of religious language and its associated form of life. What he came to realise was that religion is a significant and important language-game, and that it is important to clarify and specify its grammar and rules of use.

Much of Wittgenstein's discussion of religion hinges on talk of the Last Judgement, as if this were characteristic of religion. Maybe that is how he saw it, and it has been suggested that he saw it that way because it was characteristic of his own life — that he saw himself and his work as under judgement. However that may be, the example of the Last Judgement does bring out a central characteristic of religion in general: that its beliefs are not subject to empirical investigation. No amount of attempted positivist verification or Popperian falsification is relevant.

The simple, and some would say obvious, point here is one we have already made in the light of the positivist critique of religion. Religion is not science, and the two should not be confused; or in Wittgenstein's terms, religion and science are two different language-games with separate and distinctive 'grammars'. 'Grammar tells us what kind of object anything is. (Theology as grammar)' (PI 373). One might add 'science as grammar'. One is about the world and the other is about the things of the spirit. It is just as foolish to reject religion as bad science as it is to attempt to reinforce it by giving it scientific status. It was therefore 'ludicrous' of a certain Father O'Hara to attempt to put religious belief on a par with scientific beliefs (Wittgenstein, 1966, pp. 57-9). Religion, when treated as science, becomes superstition.

Wittgenstein likens religious belief to that of 'using a picture' which regulates your life (1966, p. 71). It is a way of looking at the world, and the religious believer who sees all his actions, indeed his whole life, as somehow under judgement, is regulated by the picture of the Last Judgement. The use of such religious pictures is acquired in precisely the same way as is the learning of the use of Kuhn's *exemplars*. That is, by learning to play the appropriate language-game, by participation in the appropriate form of life, by the acquisition of the tacit knowledge shared by the common group. Just as the would-be physicist is initiated into the art through the educational system, so, for instance, church involvement helps to supply one with the appropriate rules of the religious life.

It must be emphasised that Wittgenstein's 'Lectures on Religious Belief' were more in the nature of exploratory remarks, of thinking aloud with some of his students. They represent the tentative point his thinking had reached in 1938, a dozen years before his *Philosophical Investigations*, and as such they were incomplete and inconclusive, and they leave many questions unanswered. For instance, do Wittgenstein's religious 'pictures' correspond in any way to reality? Can we speak significantly of God's existence? Are religious beliefs reasonable or unreasonable? Can believers and non-believers contradict one another? Are there standards of rationality by which religion (and science) can be evaluated? And if religion is a form of life, does God die as the form of life becomes extinct? There was certainly more than a hint of this when Wittgenstein wrote in the *Tractatus*: 'To act ethically or believe religiously is to be in an altogether different world, which must wax and wane as a whole' (6.43).

The relativists we have discussed maintain not only that we cannot get outside all conceptual schemes to discover what 'really exists', but that each conceptual scheme confines us; we are imprisoned by our 'frameworks', 'What has to be accepted, the given, is the form of life'

(PI 226). 'This language-game is played,' and that is that (PI 654). Popper, in opposition to this relativist stance, maintains that although it is difficult, we can attempt, often with considerable success, to break out of our frameworks. Difficulties should not be construed as impossibilities. If we really were imprisoned by our conceptual frameworks there could be no genuine growth of knowledge and understanding. He pursues this theme in 'The Myth of the Framework' in terms of the alleged culture clashes which arise from what Quine calls the 'ontological relativity' of language. According to this thesis 'two languages may have built into their very grammar two different views of the stuff the world is made of, or of the world's basic structural characteristics' (Popper, 1976b, p. 37). Popper maintains that not only are most languages roughly intertranslatable, but that such culture clashes, far from prohibiting rational discussion, often prove fruitful in evoking a critical attitude. He concludes:

> It is the method of science, the method of critical discussion, which makes it possible for us to transcend not only our culturally acquired but even our inborn frameworks ... Frameworks, like langages, may be barriers; but a foreign framework, just like a foreign language, is no absolute barrier. And just as breaking through a language barrier is difficult but very much worth our while, and likely to repay our efforts not only by widening our intellectual horizon but also by offering us much enjoyment, so it is with breaking through the barrier of a framework. A breakthrough of this kind is a discovery for us, and it may be one for science (1976b, pp. 44-6)[14]

Our contention, to this point, is that the second phase of Wittgenstein's philosophy (which we do not construe in terms of a radical break with his earlier view because the central concern of both is with the limits of language and meaning) provides the intellectual roots for the *Weltanschauung* philosophy of science subsequently developed, in the manner we have outlined, by Kuhn, Feyerabend and others. Similarly, Wittgenstein's view of religion has been exploited and made far more explicit by some of his former students and by subsequent Wittgensteinian philosophers, such as Norman Malcolm, Rush Rhees and Peter Winch. D.Z. Phillips is responsible for what is probably the most extensive and uncompromising development of Wittgenstein's ideas as applied to religion, and as it is very much of a piece with the work of Malcolm, Rhees and Winch, it is to his work that we finally turn.

Religion without Explanation: D.Z. Phillips[15]

In his introduction to *Religion and Understanding* (1967), Phillips takes a 'Wittgensteinian' quotation from W.H. Watson's book *On Understanding Physics* as an indication of his whole approach to the philosophy of religion. 'Look to see what men do with things, with words and ideas, and observe their behaviour' (p. 1). Here is an echo of Kuhn's injunction with respect to the scientific enterprise, and it is not surprising that, with the same policy and programme in mind, a common view with regard to rationality is reached.

There is no doubt that in many ways this approach is a refreshing corrective to that of the traditional philosophers of religion who scrutinise the claims of religion as a branch of speculative philosophy in isolation from the actual practices of religious believers. Philosophy, as we have emphasised, is fundamental to man's behaviour and to the institutions he constructs; but philosophy reduced to academic speculation, totally divorced from the realities of life, can be an idle and stultifying pastime; scholasticism in the worst sense of the word. And because such theoretical discussions are inherently inconclusive, the rich and profoundly important religious dimensions of man's life have remained unexplored by many sceptical philosophers because they have been unable to proceed without having first established 'the existence of God'. Whatever contrary conclusions we may reach with regard to Phillips' approach to the philosophy of religion, we owe it to him to have diverted the philosopher's attention from endless and fruitless discussions about the existence of God to some of the realities of religious practice. He does not confine the application of his ideas to general, speculative principles, as indicated by the title of one of his earlier books, *Faith and Philosophical Enquiry* (1970a), but also to specific religious concerns and practices, such as *The Concept of Prayer* (1965) and *Death and Immortality* (1970b). However, although Phillips writes at times with a refreshing sensitivity and spiritual insight, a sense of philosophical perplexity remains. Instead of being clarificatory (after the alleged fashion of the Wittgensteinian method) his writing is at critical points obscure, even incoherent.

The one person, above all, who has set the terms of reference in our discussion of the various philosophies of science is David Hume. His legacy cannot be underrated, either in terms of the pervasive and widespread influence of positivism, or in terms of reactions against it. Phillips' philosophy of religion, like the *Weltanschauung* philosophy of science, represents the latter. In *Religion without Explanation* (1976), Phillips refers to the 'enormous influence' of Hume on the philosophy of religion and agrees that 'given its assumptions, Hume's attack on certain theistic arguments is entirely successful' (1976, p. ix). Phillips accepts Hume's rejection of natural theology and of the insuperable difficulty of inferring God's existence from features of the physical world. 'The whole notion of a God and another world which we can infer from the world we know is discredited' (1976, p. 21).

Phillips not only appears to accept Hume's rejection of natural theology as a philosophical method of inference from the world to God, but also his conclusions regarding the existence of God. Phillips' point here is that God is not like an object whose existence can be established (or disproved) by philosophical enquiry God is not a being among beings, an object among objects. The word 'God' is not the name of a thing which can be identified. And even if it were, the establishment of such a 'fact' would not increase our understanding or necessarily our commitment. We can no more ask general questions about the reality of God than we can ask about the reality of physical objects; each has its distinctive language-game. Just as in science it makes no sense to say that in general physical objects do not exist, so the possibility of the unreality of God does not arise *within* any religion (1965, p. 21). He enlists the support of Norman Malcolm who, in his discussion of Anselm's ontological argument (Malcolm, 1967), demonstrates that a God who is a God among existents is not the God of religious belief. Talk of God as an object imposes an alien grammar on religious faith. God, by definition, does not depend on anything else for his existence. He quotes Kierkegaard: 'God does not think, he creates; God does not exist, he is eternal' (1965, p. 81). There can be no theoretical understanding of the reality of God. Knowledge of God is not an accumulation of facts about him as is the understanding of the mechanics of a motor car. To know God is to love him. Quoting Kierkegaard again, he maintains that 'love is the spirit of God and to possess it is to walk with God' (1967, p. 74). And if to *know* God is to love him, then we cannot know God and defy him. Atheism is not a conscious rejection of God, it is simply a lack of knowledge of God that results from ignorance of the appropriate form of life. Thus belief, or unbelief, does not result from metaphysical speculation, but from playing the appropriate language-game, or being part of the appropriate form of life.

In his discussion of the religious language-game, Phillips exploits Wittgenstein's idea of the believer's use of pictures. To ask if the pictures are true is to misuse them. The picture of God is not like the picture of an aunt. We speak of God's eye seeing all, but we lapse into absurdity if we try to locate God's eyebrows. We use pictures, we believe in them, we live by them, we are judged by them: that is the nature of the game. Religious pictures must be distinguished from hypotheses, conjectures, empirical propositions, which can be put to the test. Hence to ask if religious beliefs are true is not to ask for evidence, but to ask if we can live by them: 'if a man believes death has no dominion over him . . . he will react, make decisions, different to the one who does not hold the belief' (1970b, p. 148). The pictures are not established by evidence, therefore they cannot be overthrown by evidence. A man may turn sceptical, his belief may turn to disbelief, if the picture loses its hold on him. A rival picture takes over. No contradiction between belief and unbelief is involved here, simply an exchange of pictures. The picture is not a picturesque way of saying something else. It says what it says. 'The whole weight may be in the picture' (Wittgenstein, 1966, p. 72). When the picture dies, something dies with it; the whole culture declines. Are we saying that in such circumstances God dies, if his existence derives its force and meaning from the picture? Such a conjecture is out of place, maintains Phillips; religious believers can say something now about such a loss – not that God has died, but that the people have turned their backs on God.

What Phillips, following Wittgenstein, appears to be saying here is that all our religious beliefs have an absolute character; they are in no sense hypotheses which might turn out to be false. We are held by them, captivated by the picture they present. This is why no contradiction is involved between the believer and the unbeliever; they just have different pictures, different perspectives. The atheist who denies the existence of God does not contradict the theist who puts his trust in God. He is 'rejecting a whole mode of discourse' rather than 'expressing an opposite view within the same one' (1967, p. 183). Thus our beliefs are not so much propositions about reality, they are part and parcel of a way of life which involves such things as prayer, worship, praise and penitence.

Phillips insists that there can be no general justification of religious belief, giving religion a sure foundation. We cannot stand outside religion to judge it. Such an attitude is based upon a mistaken epistemology, claims Phillips, which assumes a standard use of language with a constancy and transparency of meanings, and a criterion of justification or of truth which is inaccessible to us. The fact is that there is no one paradigm of rationality to which all human activity must conform. Therefore religion (like natural science) can only be described and not evaluated objectively in terms of truth and falsity. The internal theologies of different religions give them their meaning. So the task of the philosopher of religion is to clarify the grammar of religious beliefs and not to take sides for or against those beliefs. 'What has to be accepted, the given, is the form of life' (PI 226).

Attempts to evaluate the rationality of religious beliefs in terms prescribed by Humean empiricism are therefore fundamentally mistaken; they arise from a confusion of two distinct language-games. Crucial to understanding Phillips' position here is his rejection of metaphysical speculative thought as a 'confusion about the grammar of our language' because 'the metaphysician attempts to go beyond the limits of what can be asked. He wants to ask what cannot be asked' (1976, pp. 107 and 111). But, following Wittgenstein, he insists that metaphysics, like the mystical, and like magic and religion, 'makes a deep impression on us'; it is an essential part of our lives which needs to be acknowledged and explored. Therefore, instead of being drawn into a discussion concerning the cognitive nature of religion, we would more profitably engage in an enquiry 'into what it is that makes us want to speak in this way' (1976, p. 105). The absolute character of our beliefs endows them with a sense of ultimacy beyond which we cannot venture. In this sense, religious belief is not a rational affair, but neither is it irrational, maintains Phillips, it is non-rational. This is why the philosopher's task is descriptive and clarificatory. Thus Phillips turns our attention to the religious form of life, of which its language is but a part, and instead of asking, 'Is there such a being that . . . ?' we should ask, 'What is the grammar of our idea of God, what can be said about him?' (1965, p. 81). 'Discovering that there is a God is not like establishing that something is the case within a universe of discourse with which we are already familiar, but the discovery that there *is* such a universe of discourse' (1967, p. 69).

We are reminded that it was the failure to reduce the symbolic generalisations and metaphysical principles, by correspondence rules, to empirical phenomena, that led to the collapse of the Received View of science, and that it was this which led Kuhn to turn to the scientific community which is characterised by these shared metaphysical principles. Kuhn's equivocation about the rationality of science is akin to Phillips' resort to the notion of non-rationality. The legacy of both Hume and Wittgenstein is evident in each enterprise; both religion and science are secure in entrenched positions and rendered invulnerable to the anti-metaphysical attacks of positivism.

Having dismissed the metaphysical character of religious belief as resulting from a confusion of 'grammar', or a misuse of language, Phillips directs our attention to what he detects as the central feature of the religious form of life; it is above all *expressive* — of man's hopes and fears and aspirations. Thus he writes:

> The magical and religious beliefs and practices are not the confused outcome of deep problems and emotions, but are themselves expressions of what went deep in people's lives. That a man's misfortunes are said by him to be due to his dishonouring the ghosts of slain warriors is itself the form that depth takes here; it is an expression of what the dead mean to him and to the people amongst whom he lives. That a man says that God cares for him in all things is the expression of the terms in which he meets and makes sense of the contingencies of life. (1976, p. 114)

The man who 'puts his trust in God', and who speaks of 'God's care for us' is not articulating a belief in a transcendent, supernatural being or power: rather he is giving expression to a way of coping with the vicissitudes of life. He quotes Rush Rhees' remark, 'we can only *describe* and say, human life is like that' (1976, pp. 112-20). Prayer does not imply a celestial telephone link with the Almighty, it expresses a state of being, a state of the soul. The prayer speaks for itself (1965, p. 109). Likewise, talk of the reality of the dead does not imply the existence of another world beyond human experience; it is rather expressive of a way of facing and coping with the death of one's loved ones (1965, pp. 122f). Immortality has nothing to do with the survival of earthly death, or of future reward or punishment, but with living according to God's commands now; this is life eternal which death cannot reach. The moral of Phillips' *Death and Immortality* appears to be 'when you are dead you are dead, but live a good life now because it is of eternal significance'. He takes up Wittgenstein's reference to the Last Judgement. Again, the language is not descriptive of a future event, but rather it is valuational, expressive of a picture which constantly admonishes us, or which suggests that we are answerable for all our actions (1970b, pp. 145f).

Phillips' account of religion is reminiscent of the emotive theory of ethics which some positivists (Ayer, for example, in *Language, Truth and Logic*) resorted to in their attempt to account for moral judgements which appeared to be rendered meaningless by the strictures of the Verification Principle. The metaphysical interpretation of religion is forbidden on precisely the same Humean grounds, and so its characteristic utterances are construed entirely in terms of their profound expressiveness of the deeply rooted emotions of men, and the role they play in human life. And so Phillips, following Wittgenstein, sees the task of the philosophy of religion as an attempt to reveal and make explicit this expressiveness which is obscured by the metaphysical interpretation of the traditional philosophy of religion.

This at once points to a curious contradiction contained in Phillips' philosophy of religion. Although, consistent with his whole philosophy, he views science, like religion, as a form of life, or as a language-game with its distinctive grammar, he also appears to acknowledge the Humean, positivistic view of science with its emphasis on justification by empirical verification, and its denial of cognitive significance to non-empirical, metaphysical assertions. He correctly diagnoses that it is this view of science which is lethal to religion, but instead of examining the philosophy of science which contains such a devastating critique of religion, he encapsulates religion in the form of life philosophy, thus insulating it from Humean criticism by putting it beyond reason. But the whole point of the form of life approach is that it denies the very principles on which positivism is founded. For a central tenet of positivism is that there can only be *one* 'grammar' of truth, that which involves some degree of empirical verification. To admit, as for example Phillips does, that there are different 'grammars' or levels of truth is contrary to the whole of the logical empiricist enterprise. Our contention is that the *Weltanschauung* philosophy of science is the counterpart to the Wittgensteinian philosophy of religion, as developed by Phillips and others, and that both are reactions to the presumptions of positivism.

Meaning and Existence

If our contention is correct, then Wittgensteinian 'fideism' is subject to the same flaws as the *Weltanschauung* philosophy of science, and both are forms of irrationalism. The two main criticisms concerning meaning and objectivity, which we levelled at the *Weltanschauung* philosophy of science, are equally applicable to the fideistic approach to religion which we have attempted to outline.

Phillips in particular takes Wittgenstein's remark 'Grammar tells us what kind of object anything is. (Theology as grammar)' as a licence for concern with *meaning* rather than with truth or existence. This may well be a misconstruction of Wittgenstein who could be presumed to have meant that theology tells us what sort of object God is, and so not ruled out discussion of the objects of religious belief and practice. Renford Bambrough comments:

To know what kind of object a thing is is to be equipped or partially equipped to determine whether there are any objects of the kind in question, and what properties and relations, among those that are recognised in our 'grammatical' understanding to be relevant, actually characterize any such objects of that kind as may actually exist. (1977, p. 17)

However this may be, Phillips is correctly following Wittgenstein in insisting that religious beliefs are on a totally different plane from scientific, or everyday, ones. Their meanings are incommensurable with those of science. But the question arises, how do we *know* what people mean when they make (religious) assertions? Phillips, like Kuhn with respect to science, attempts to draw our attention to the realities of religious practice, and he is doubtless correct in drawing our attention to the *expressive* nature of religious beliefs, but why should they not be literal and descriptive as well? Why should they not be both? H. A. Williams, for example, in *True Resurrection* (1972), draws the Christian's attention to ways in which we can understand the resurrection of Christ by living the resurrection life now, in this life. But he does not thereby ignore the reality of the resurrection of Christ as an historical event, problematic though it may be. Correspondingly, the Christian who takes the idea of the Last Judgement not simply as a metaphor may not necessarily envisage Michelangelo's powerful representation in the Sistine Chapel; he could still believe in some form of ultimate individual assessment. It is surely more likely that the *expressive* use of religious language, that is expressive of how one copes with the contingencies of life, arises from the literal belief in an almighty, loving God. On what grounds does Phillips assume that 'the grammar of "belief" and "truth" is not the same as in the case of empirical propositions or the prediction of future events'? (1976, pp. 142-3). It seems that Wittgenstein is here legislating for a form of life rather than describing it, and, like Kuhn, under the guise of telling us how scientists do in fact behave, is prescribing how scientists ought to behave.

The plain fact of the matter is that scientists do believe in an objective world which is the subject of their investigations, and religious people likewise believe in the objective reality of God. Ontology is an essential strand in the meaning of both scientific and religious discourse, the elimination of which makes a nonsense of the activities and utterances associated with science and religion. If we pay attention to the living language of religion we cannot avoid its 'other worldly' references. Whether or not metaphysics 'tries to say what cannot be said' in its reference to God's existence, or to another world, the fact is that people do speak in this way and it is a fair assumption that they mean what they say — in Wittgensteinian terms, the language says what it says. Phillips rightly grounds religion in things such as prayer, worship, reverence and devotion, but these activities only make sense if they have an objective and purposeful reference. Certainly Christianity, as an historical religion, refers to the God of Abraham and Isaac, of Saint Paul and the Evangelists; and central to its meaning is the incarnation of God in Christ at a particular place and time. We have referred to the attempts of Braithwaite and van Buren and others to side-step these issues, because of their problem with metaphysical assertions, but the Wittgensteinian descriptive philosophy cannot deny that the 'saints' down the Christian centuries have staked their lives on such a faith. Nor can it be denied that there is a long and distinguished tradition of intellectual enquiry, conducted by such giants as Augustine, Anselm and Aquinas and more recently by Barth and contemporary religious thinkers, into the content and meaning of the Christian revelation. We may argue that they came to the wrong conclusions, but not that they did not know what they were about, or that they somehow did not really understand the rules of their language-game. The fact is that Phillips, and others following Wittgenstein, are wrong in saying that 'philosophy leaves everything as it is'. Philosophical enquiry not only changes people's lives, but it also succeeds in solving some philosophical problems. (Wittgenstein's own outstanding contribution to the philosophy of mind in the *Philosophical Investigations* is held by many to be such an example.) Phillips' account of religion allows no place for polemic, for preaching and apologetic, for mission and conversion, and certainly no place for rational argument. He leaves no place for doubt and scepticism and questioning amongst religious believers. Belief, for Phillips, involves total commitment, unshakeable conviction. To acknowledge God is to worship him. But again this is contrary to the evidence. It is not self-contradictory to acknowledge God and repudiate him completely. As Roger Trigg has nicely put it, 'This is presumably the position the Devil holds in Christian theology' (1973, p. 41).

Most Christians are faced with the moral dilemma posed by Saint Paul: 'For the good that I would I do not: but the evil which I would not, that I do' (Romans 7.19, AV). We know God and his will for us, but we so often do not do it. It might be unreasonable, or irrational, to know God's commandments and yet not live by them, but again, this is the story of the religious believer's life. Further it is a commonplace that religious belief can be shaken or reinforced, faith and doubt can be generated by argument and evidence. Faith and doubt often go hand in hand. 'Lord, I believe; help thou mine unbelief' (Mark 9.24, AV). And far from the atheist and the theist playing incommensurable language-games, atheists *do* understand (at least in some degree) the claims of religion but they reject them as false. Presumably this is the position of the reformed Ayer. If the Wittgensteinian says that this is not true religion, we can counter by saying that he does not understand the religious form of life!

In the same way that the *Weltanschauung* philosophy of Kuhn and Feyerabend deprives science of any sense of objective reality, so the Wittgensteinian philosophy of religion denies that talk of the objective reality of God can have any sense. But belief in God implies that he existed when men and their language-games did not. Although talk of God, as the creator of the world without men, is part of the religious language-game now, there is an objective reference to a state of affairs in which (to use Vanstone's phrase) the response to God's creative love is confined to that of nature, and is not the response of human freedom or recognition.

In his *A Realist Theory of Science* (1978), Roy Bhaskar attempts to hold in balance the fact that knowledge is a social product, dependent on the men who produce it, and that knowledge is *of* things which are not produced by men at all. He writes, 'If men ceased to exist sound would continue to travel and heavy bodies fall to the earth in exactly the same way, though *ex hypothesi* there would be no-one to know it.' These *intransitive objects of knowledge*, as Bhaskar calls them,

> are in general invariant to our knowledge of them: they are the real things and structures, mechanisms and processes, events and possi-bilities of the world; and for the most part they are quite indepen-dent of us. They are not unknowable, because as a matter of fact quite a bit is known about them . . . but neither are they in any way dependent upon our knowledge, let alone our perception, of them. They are the intransitive, science-independent, objects of scientific discovery and investigation. (1978, pp. 21-2)

> Science is the systematic attempt to express in thought the struc-tures and ways of acting of things that exist and act independently of thought . . . the question of what is capable of independent exist-ence must be distinguished from the question of what must be the case for us to know that something is capable of independent exist-ence . . . There could be a world without men; but there could not

be knowledge without antecedents. (1978, p. 250)

Einstein and Infeld write in similar vein:

> Science is not just a collection of laws, a catalogue of unrelated facts. It is a creation of the human mind, with its freely invented ideas and concepts. Physical theories try to form a picture of reality and to establish its connection with the wide world of sense impressions. Thus the only justification for our mental structures is whether and in what way our theories form such a link. (1961, p. 294)

It is precisely this realism which is denied by the proponents of the *Weltanschauung* philosophy of science. It contrasts sharply with Feyerabend's assertion that 'objective reality has been found to be a metaphysical mistake', and that realism is simply a ploy used to persuade others to accept our beliefs (1981, Vol. I, p. xii). The epistemology that generates Feyerabend's subjective idealism is precisely the same as that which leads to Phillips' denial of objective reference in religious language. Each represents a reaction to the justificatory and authoritarian philosophy of positivism, and each is intellectually rooted in the later phase of Wittgenstein's thought.[16]

Wittgensteinian fideism represents the philosophical counterpart, and may well have added impetus, to the strong current tendency in Christianity to play down credal statements and doctrinal formulations, and to locate the centre of religion rather in 'a personal relationship with Christ'. But such a Christ-centred faith makes little sense without the presuppositions of objective belief, however loosely formulated in credal statements. The intimate personal terms of 'trust', 'guidance', 'reliance' and 'surrender' all imply an objective reference, and need grounding in a traditional rational philosophical and theological manner.

James Richmond, in a review of Phillips' *Religion without Explanation*, adds a further dimension to this criticism when he suggests that it offers

> a welcome escape to those who are frustratingly tired of what sometimes appears as the intolerable burden placed upon them by the unflagging demands of the philosophy of religion, demands which have to be responded to in every generation. How wonderful, we have heard it said, to be released from the intellectual burdens of the

existence of God, the problem of evil, the relation of God to the spatio-temporal world, the question of verification, the nature of theological ethics, the notion of life after death, and the rest. And in each generation we see the results of this escape being effected, in arid biblicism, in one form or other of 'neo-orthodoxy', in submission to various forms of external authority. Now Phillips pleads for that change of direction and radical turnabout in contemporary philosophy of religion programmed in his book. The danger is that many of the intellectually frustrated, to be sure very much less able and gifted than he, will respond positively to his plea. But before doing so, it should be made clear to them that a philosophical exploration of religion should shed light upon it rather than darkness, and that their theological frustrations and quandaries might be worsened if they were to exchange one set of problems for others which turn out to be more rather than less intractable. (1980, p. 43)

What is the attraction of the relativism we have described, in both science and religion? Why should men turn from the objective pursuit of truth to a Protagoraean individual or collective subjectivism? To this question we will turn in our conclusion, but first we must attempt to gather up the threads concerning objectivity, truth and realism which are at the heart of the debate.

The Slippery Subject of Truth

One of the central issues in our threefold enquiry into the foundations of science is that of truth. The positivist attempt to link verification with truth foundered once it was admitted that all our knowledge is fallible. The scepticism generated by the failure of this positivist, Received View of science leads, via meaning analysis, to a relativism which denies truth in any objective sense.

It is when we examine the extreme relativist case, presented by Feyerabend, that we discern its implausibility. Feyerabend derides the urge to seek truth: his central concern is with individual human happiness which, he maintains, is best enhanced by a social and intellectual anarchy in which people are free to follow their own inclinations and fancies and whims. John Krige aptly summarises Feyerabend's flippant Dadaism in terms of a belief that

life will only be worthwhile when we stop taking it too seriously, and free ourselves from a puritanical and dedicated search for 'truth' and 'justice' . . . Reasoning is simply a game which the Dadaist plays preferably with those who are naive enough to believe that we can get closer to the truth by attending to the content of arguments and by discussion. (Krige, 1980, p. 109)

Such a position is self-defeating for it invites the retort, 'Why bother to take Feyerabend's arguments seriously?' Presumably Feyerabend is advocating a position which he thinks is true and worth while – the very values he derides as puritanical.

Popper rejects both the positivist identification of truth with verification, and the relativist dismissal of truth as a chimera, and he sets up truth as a metaphysical goal or a regulative principle. However hard we try, we cannot eradicate the concept of truth. It is biologically and intellectually embedded in human aspirations and endeavours, and there is no way that 'truth' can be taken out of our vocabulary. All human enquiry, from the rigours of logic and mathematics, to scientific theorising, to metaphysical speculation, to theological investigation, assumes some evaluative principle of truth. But, Popper emphasises, there can be no criterion of truth. We cannot step outside the world and identify truth in an absolute sense. We have no absolute yardstick to which our explanatory theories and conjectures, whether of science or metaphysics, must conform. This is why human knowledge is fallible and we can never know how close to the truth we have got.

However, truth is not an independent metaphysical entity. It only makes sense as a part of a (Hegelian) dialectic; that is, in opposition to falsity. Truth and falsity are two sides of the same coin. Truth also implies realism, whether of the workings of the physical world or of the nature of God and his relationship with the world. These are the claims and working assumptions of both scientists and theologians.

Nevertheless, reality is not an abstract Platonic realm totally independent of our comprehension of it. This is where descriptive epistemology comes in; that is, the link between science and philosophy. Wittgenstein's talking lion can never comprehend the world in the way that Wittgenstein (if we may resurrect him!) can, because of biological structures and limitations. However, the lion and Wittgenstein occupy the same world, as do the amoeba and Einstein (if we may resurrect him too!). It is this real world that man explores with his theories, and in so doing achieves better and better explanations. Maybe we will one day encounter extra-terrestrial beings of superior intelligence who comprehend the world (and its Maker) in a fuller and more realistic sense than we do. It is certainly true that amongst human beings there exist wide ranges of intelligence and cognitive ability; few people in the world today are able to see the world in Einsteinian, relativistic terms, or to conceive of gravity in geometrical terms of curved space. Can we therefore extrapolate this idea of increasing comprehension to some ultimate cosmologically omniscient being, say God, who would have a complete list of all true statements about the universe, and presumably about himself? In other words, is there some cognitive ideal of total knowledge, a full account of things as they really are?

Gonzalo Munévar, in his book *Radical Knowledge*, suggests that this is what Popper's notion of absolute truth implies. But on Popper's own account this cannot be so. Given his descriptive epistemology with respect to the evolution and functions of human language, language is only an approximate, and often misleading, affair: although, as Russell said, it is all we have, and in a strange way it serves us pretty well. But because of its approximate and flexible character it makes no sense to speak of the complete class of all true theories: that is, of theories that precisely and unequivocally match the world. Popper's point about absolute truth is a logical one: in practice it can never work out. Not even our omniscient being could provide such an absolute and final description. Human language is still human language, even in the mouth of God.[17]

Munévar, however, does develop an enlightening thesis concerning reality and truth and objectivity. Pursuing Popper's point that knowledge comes from an interaction between an organism and its environment, he suggests that perceptions and theories and cognitive appraisals are relative to frames of reference in the same way that mass, distance and time are relative to frames of reference in relativity theory. Thus there is no preferred frame of reference, no absolute reality, no *one* way the world is. But it does not follow that all frames of reference are equally good: presumably Wittgenstein had a better view than his lion; a man comprehends the world in a fuller way than his dog. Such a theory of scientific relativism, like its progenitor perceptual relativism, does not deny the reality of the world. There *is* a world and there are descriptions and explanations of it that rank in terms of verisimilitude.

This brief, and final, encounter with the slippery subject of truth is but a tentative contribution to current debate, but it is relevant to our thesis because if objectivity, truth and realism are dismissed from the scientific enterprise, then by the same token they cannot rank in the vocabulary of theological debate and religious enquiry.

186   *The Retreat to Irrationality*

Notes

1. Reichenbach (1938, pp. 6-7) distinguished between the *context of discovery* and the *context of justification*, insisting that epistemology is concerned only with the latter; the former being the concern of psychology and history. John Krige (1980) contests this 'rigid distinction between psychology and logic, and between discovery and justification' (p. 214), maintaining that both are open to rational assessment.

2. See 'Normal Science and its Dangers' (Popper, 1970).

3. All references in this section are to Kuhn unless otherwise indicated.

4. Both Kuhn and Popper agree that scientific change is revolutionary, but Popper, unlike Kuhn, insists that revolutions are, or ought to be, a permanent feature of science. See 'The Rationality of Scientific Revolutions' (Popper, 1981).

5. See the discussion in Kuhn's paper 'Second Thoughts on Paradigms' (1977a) and Suppe's 'Afterword' (1977).

6. All references in this section are to Feyerabend unless otherwise indicated.

7. This is beautifully demonstrated with great clarity and simplicity by Einstein and Infeld in *The Evolution of Physics* (1961) in which they trace the evolution of our understanding of the physical world, linking early mechanistic accounts with developments in field and relativity theory and quantum discontinuity.

Hermann Bondi, in *Relativity and Common Sense* (1962), as the title of his book implies, derives Special Relativity from simple non-relativistic notions.

John Krige examines the relationship between continuity and discontinuity, and their rational appraisal, in his book *Science, Revolution and Discontinuity* (1980). He traces the development of these ideas from Popper to Lakatos, Kuhn and Feyerabend, and although critical of the distinction between discovery and justification, he comes close to the Popperian position we have attempted to outline. He concludes:

It is to Popper's lasting credit that he saw growth as requiring opposition, and that he attempted to develop a theory of rational criticism. It is also to his credit that, having a sufficiently astute sense of history, he realised that an entire established order cannot be rejected in one fell swoop, and that we cannot start from scratch. (p. 214)

Einstein was well aware of this, ever conscious of his debt to Newton, the giant on whose shoulders he stood (see *Einstein*, by Banesh Hoffmann (1975), pp. 139-42 and 247-8).

None of this should imply that scientific theories are not revolutionary and do not involve profound conceptual changes, but *that radical theory changes are not incommensurable*.

8. Popper is distinguished by his hostility to both phases of Wittgenstein's philosophy (although he had more time for the first than the second) and he produced a philosophy of science and a theory of knowledge which avoid both extremes.

9. Wittgenstein himself never claimed that all specifically identifiable disciplines and activities in which people engaged were separate language-games, each with its own set of rules or grammar. But many of his followers made the assumption, and consequently two generations of philosophers wrote books with titles such as: *The Vocabulary of Politics, The Language of Morals, The Logic of the Social Sciences, The Logic of the Sciences, The Language of Education, The*

Logic of Religious Language, The Language of Christian Belief and so on *ad nauseam* (see Bartley, 1974, p. 121).

10. The rererences to Wittgenstein's *Philosophical Investigations* are given by PI followed by a number referring to the paragraph.

11. This extraordinary statement, that human understanding is about words and not the world, is not in keeping with the general tone of Hudson's writings, but it indicates the ease with which many post-Wittgensteinian philosophers fell into the language-game trap.

12. Russell stated with regard to the later philosophy of Wittgenstein:

> Its positive doctrines seem to me trivial and its negative doctrines unfounded. I have not found in Wittgenstein's *Philosophical Investigations* anything that seemed to me interesting and I do not understand why a whole school finds important wisdom in its pages . . . the later Wittgenstein seems to have grown tired of serious thinking and to have invented a doctrine which would make such an activity unnecessary. (1975, pp. 160-1)

Popper, using one of Wittgenstein's oft-quoted examples, suggests that it is Wittgenstein himself who is buzzing about in the fly bottle, unable to escape (1971, Ch. 8).

Peter Munz commented that Wittgenstein's later philosophy 'taught young men to criticise formulations rather than ideas' (1964, p. 88).

13. Hudson argues this at length (1975, Ch. 3).

14. Denis Nineham's *The Use and Abuse of the Bible* (1976) provides an interesting example of extreme cultural relativism. His concern with the social and cultural framework within which Biblical ideas emerged parallels Kuhn's concern in *The Structure of Scientific Revolutions*. Like Kuhn, he refers to paradigm changes from one 'totality of experience' to another. Nineham's book could in many ways be construed as an acceptance of Kuhn's challenge 'for similar and, above all, for comparative study of the corresponding communities in other fields' (Kuhn, 1970a, p. 209). John Barton, in a critical review of Nineham's book, points to the central paradox of such relativist approaches, and he echoes Popper when he writes:

> No-one can escape from his cultural background; yet, if this is so, cultural change cannot occur, which is obviously false. If, on the other hand, cultural change does occur, then men must (at least on occasion) be able to escape partially from their own culture, and thoroughgoing relativism must be abandoned. [He concludes:] Modern theology cannot ignore the relativist analysis; but it would do well to remember that very simple solutions to the problems of human life and culture are rarely very satisfactory, and to refuse to be forced into defending its beliefs as if they were to be understood as statements of 'hard facts' like the propositions of nineteenth century science. (Barton, 1979, pp. 196 and 199)

15. All references in this section are to D.Z. Phillips unless otherwise indicated.

16. Don Cupitt, as we noted in Chapter 1, like Phillips, uncritically accepts the positivistic, Received View of science and later courts the relativist position. Assuming, without question, the instrumentalist thesis, he claims that scientific theories are simply 'interpretive tools' and that ' "reality" changes as theory changes' (1983, p. 239). He refers disparagingly to the 'fetishism of objectivity' (p. 334) and boldly asserts that 'what faith is shows what God is' (p. 332). Here is further evidence that relativism is rooted in positivism.

17. This is why Biblical fundamentalism is such a pernicious and dangerous doctrine. The Biblical fundamentalist does not understand the nature and limitations of human language.

4 SCIENCE, RELIGION AND RATIONALITY

Philosophy is never finished. It is, by its very nature, an inconclusive subject. This is because philosophy is such a wide and all-embracing affair, co-extensive with man's curiosity and enquiry into the nature of the cosmos and his own significance within it. The business of philosophy will be ever incomplete as long as there is a world containing enquiring minds. And the enquiry has ever been twofold (although in some respects it can be viewed as one), directed at God and the world. We noted in our introduction how, since the rise of modern science in the seventeenth century, the enquiry has been increasingly concerned with the physical world; the world which we now investigate at so many levels, from the human sciences of psychology and biology, through its chemical constitution, to the theory of matter at its puzzling quantum level and on its breathtaking cosmological scale. But God will not go away; even in an age in which voices increasingly, and almost triumphantly, speak of 'the death of God', the deep and disturbing questions of religion persist. The Alpha and Omega of physics, spelled out in the spectacular terms of the 'Big Bang' theory of the creation of the universe, have not displaced the Alpha and Omega of religious enquiry.

Bertrand Russell, one of the most penetrating and imaginative of the enquiring minds of the twentieth century, wrote:

> That man is the product of causes which had no prevision of the end they were achieving; that his origin, his growth, his hopes and fears, his loves and beliefs, are but the outcome of accidental collocations of atoms; that no fire, no heroism, no intensity of thought and feeling, can preserve an individual life beyond the grave; that all the labours of the ages, all the devotion, the inspiration, all the noonday brightness of human genius, are destined to extinction in the vast death of the solar system, and that the whole temple of man's achievement must inevitably be buried beneath the debris of a universe in ruins — all these things, if not quite beyond dispute, are yet so nearly certain, that no philosophy which rejects them can hope to stand. Only within the scaffolding of these truths, only on the firm foundation of unyielding despair, can the soul's habitation henceforth be safely built. (1929, pp. 47f)

Some nineteen hundred years before Russell spelled out this moving obituary of man and his achievements, an old man, pondering much the same mighty matter at the evening of his life on the Isle of Patmos, wrote:

And I saw a new heaven and a new earth: for the first heaven and the first earth were passed away; and there was no more sea. And I John saw the holy city, new Jerusalem, coming down from God out of heaven, prepared as a bride adorned for her husband. And I heard a great voice out of heaven saying, Behold, the tabernacle of God is with men, and he will dwell with them, and they shall be his people, and God himself shall be with them, and be their God. And God shall wipe away all tears from their eyes; and there shall be no more death, neither sorrow, nor crying, neither shall there be any more pain: for the former things are passed away. And he that sat upon the throne said, Behold, I make all things new. And he said unto me, Write: for these words are true and faithful. And he said unto me, It is done. I am Alpha and Omega, the beginning and the end. I will give unto him that is athirst of the fountain of the water of life freely. He that overcometh shall inherit all things; and I will be his God, and he shall be my son. (Revelation 21. 1-7, AV)

The 'Passionate Sceptic' and the 'Beloved Disciple' were, in their different ways, addressing the same great questions, and it is our contention that although the branches of religion and science have grown far apart, the latter in recent years far outstripping the former, not only have they sprung from the common root of human enquiry, but the *philosophical* problems associated with the one have much in common with the *philosophical* problems associated with the other. And the philosophical probems will not go away.

We have examined three important and influential approaches to the common philosophical roots. And although each philosophical theory appears to extinguish its predecessor (such is the arrogance of philosophers!), they are a part of a continuing and connected attempt to give an account of the extraordinary phenomenon of human knowledge, and the enquiry into meaning and truth. In retrospect one can see that the strength of each philosophical approach is also its weakness. In their zeal to develop their arguments and offer solutions to the problems, philosophers often tend to over-emphasise one important strand of thought at the expense, and to the exclusion, of others.

Thus positivism, in its attempt to eradicate the dubious metaphysical elements from scientific theorising, ends up by disposing of the very principles on which science rests, and so destroys itself. Wittgenstein, in rejecting his earlier referential thesis of language, confines philosophy to a descriptive study of how language is used, and ends up in the extraordinary position of having purged philosophy of its central characteristic of problem solving. Popper's uncompromising fallibilism and denial that we can ever lay claim to truth can lead, if unchecked, to an irrational scepticism. Kuhn's instructive attempt to identify the characteristics of the scientific community, coupled with Feyerabend's analysis of meaning (and his refreshing reminders about the contemporary tendency to worship science), leads to a cultural and conceptual relativism and subjective idealism which are entirely at odds with the alleged objectivity of human enquiry. And in tracing the influence of these ideas in religious thinking we have seen how theologians and philosophers of religion, in uncritically subscribing to their conclusions, have embraced various forms of reductionism, scepticism and relativism almost to the point of atheism. It is precisely for this reason that the philosopher of religion needs to be thoroughly acquainted with the philosophy of science.

If the philosophy of science is important for the philosophy of religion, a knowledge of science is important for both. Indeed, we would agree with both Russell and Popper that philosophers in general should be well acquainted with science, and that it is the almost total disregard of scientific knowledge that has been responsible for what Russell has described as the blind alley of meaning analysis and the 'Philosophy-Without-Tears-School' whose Bible is Fowler's *Modern English Usage* (1975, p. 170).[1] But Russell insists that there are no quick, easy routes in philosophy, no special wisdom is required, 'only the power of patient thought' (quoted by Alan Wood, 1975, p. 202). And the power of patient thought, informed by a deep understanding and knowledge of science, should be the hallmark of the philosopher of religion as much as it should be of the philosopher of science. Whilst we are quoting Russell, his further comment that 'Philosophy . . . is something intermediate between theology and science . . . a No Man's Land' (1946, p. 10) supports our contention that the link between science and theology is philosophical, and that philosophy is important to both. Of course there are many scientists who pursue their work with little overt knowledge of philosophy, and some who would maintain that science has no need of philosophy. This may well be true of routine or 'hack' science, but it is certainly not true of science in the 'heroic sense' referred to by Popper. One has only to point to Einstein, probably the greatest scientist since Newton, who has a volume devoted to his ideas in *The Library of Living Philosophers*, entitled *Albert Einstein: Philosopher-Scientist* (Schilpp, 1949). For Einstein, physics and philosophy were one.

If philosophy is in general important for religion and for science, and is the link between the two, the particular aspect of philosophical enquiry that we have identified as a recurring theme is that of *rationality*. And it is our contention that science and religion are not necessarily two separate cultures or *Weltanschauungen*, one rational and the other irrational, but one. The scientist and the religious enquirer, far from being engaged in incompatible and opposed activities, are both trying to understand the world and man's place in it. Each uses irrational, or prerational, faculties of creative imagination, of instinct and intuition, but subject to critical control and rational appraisal. It is clear from the preceding remark that our preference is for the Popperian identification of rationality with criticism. The positivist attempt to characterise rationality in terms of empirical verification and truth is as lethal to science as it is to religion, because of the Humean strictures on inductive inference and the impossibility of proof and certainty. We also noted that the attenuated version, linking rationality with high probability, is equally fated. Attempts to avoid the problem of rationality by construing science and religion in terms of *Weltanschauungen* analyses are not acceptable because they prohibit the central concerns of both scientific and religious enquiry: reality, objectivity and truth. These three *bêtes noires* of the relativist are not to be construed in terms of proof, certainty and indubitable authority. Rather, they rest on the common-sense assumption that there is a real world to probe with our scientific theories; that our investigations, not despite but because of our perceptual and mental apparatus, do have an objective reference; and that truth remains the guiding principle. One of the miracles of evolution is that man, who has evolved *from within* the world, and is an *integral part* of the world, has developed a self-conscious mind which enables him to view the world objectively and reflect on its origins and its destiny. The mystical reflections of St John the Divine and Bertrand Russell are inexplicable, but they witness to this extraordinary feature of human life.

Although common folk (as opposed to philosophers) do not question the existence of the physical world as an objective reality, they do question the existence and reality of God. Despite the scepticism generated by positivism, the question will not go away; and despite the attempts of the Wittgensteinians the question remains one asked in terms of reality, objectivity and truth — 'simply true, not "true" with some special grammar' (Mackie, 1982, p. 228). If it is to be protested by Humean and Wittgensteinian philosophers of religion that these terms only have reference, or make sense, in empirical terms, we are justified in asking who has legislated for such a restricted use? (Back to positivism again.) One of the forms of life within which these terms are used is the religious form of life. 'God-talk', involving such terms as truth, objectivity and reality, is a language-game that is played. The Wittgensteinians have simply not made out their case when they assert that those who use religious language intend it as purely expressive, with no ontological significance.

Although present-day philosophers are more prone to challenge the credentials of religion rather than of science,[2] they are prepared to enter the lists of rational debate on questions relating to theism. Amongst a number of important contributions recently made in this field are Richard Swinburne's *The Coherence of Theism* (1977) and *The Existence of God* (1979), and J.L. Mackie's *The Miracle of Theism* (1982), and it is significant that both of these authors are well versed in the philosophy of science. As Mackie says in his introduction: 'It is my view that the question whether there is or is not a god can and should be discussed rationally and reasonably, and that such discussion can be rewarding, in that it can yield definite results.' He maintains that the question is a genuine and meaningful one and 'too important for us to take sides about it casually or arbitrarily' (1982, p. 2). He accepts Swinburne's account of the central doctrines of theism that there is a god who is

a person without a body (i.e. a spirit), present everywhere, the creator and sustainer of the universe, a free agent, able to do everything (i.e. omnipotent), knowing all things, perfectly good, a source of moral obligation, immutable, eternal, a necessary being, holy and worthy of worship. (1977, p. 1)

The arguments involve deduction, coherence and internal consistency, and appeal to evidence and experience. Because the arguments are not empirically testable and fall on the metaphysical side of Popper's criterion of demarcation, it does not follow that they cannot be rationally and critically discussed, nor does it follow that empirical evidence is irrelevant. At this point we are well reminded of Popper's contention that as long as metaphysical questions can be rationally criticised we should take seriously their implicit claim to be considered as tentatively true. Such argumentation is the lifeblood of philosophy. The tentative approach to truth (or falsity) is the linchpin of Popper's epistemology, and according to his criterion of demarcation it is equally applicable to the rationally criticisable theories and conjectures of science and religion.

Russell perceived that the Cartesian attempt to argue deductively from indubitable premises to secure conclusions was doomed to failure, and he abandoned his attempt to establish certainty and truth in science as well as in religion. It was Russell who wrote, 'the demand for certainty is one which is natural to man, but is nevertheless an intellectual vice' (quoted by Wood, 1975, p. 199).

Rather than argue from premises to conclusions, much of Russell's argument proceeded in the reverse direction. He attempted to establish premises by arguing from the conclusion or the consequences. Alan Wood writes, 'He saw the role of a philosopher as analogous to that of a detective in a detective story; he had to start from results and work backwards by analysing the evidence' (1975, p. 196). This is of particular interest, referring as it does to one of the century's great philosophers of science, because it highlights an important feature of the nature of apologetic in the philosophy of religion. Thus philosophers who are concerned solely with the logical minutiae of 'the existence of God' tend to overlook the enlightening attempts at exploring the *nature* of God rather than his existence. Likewise in science, although philosophers discuss questions relating to the existence of the world, scientists direct their efforts at investigating the *nature* of the world. (We must not confuse this point with Phillips' contention, that our task is to discuss the nature of *religious language* in its embedded form of life, because a discussion about the nature of God implies an ontology which he maintains is illegitimate.) We have outlined Vanstone's contribution in this respect and we refer to a similar effort by J.A. Baker. In *The Foolishness of God* (1970), Baker argues: given the world and human experience as we know it, what kind of God makes sense? The argument is subtle and wide-ranging, and it is perhaps not surprising that as a Christian theologian, Baker arrives at the conclusion that his investigation leads him to the Christian conception of God which although 'foolishness to the Greeks' (the philosophical sceptic of today?) is the best, in fact the only fit.

The fact that such arguments are logically inconclusive need not turn us into irresolute sceptics or relativists. Again it was Russell who perceived that 'Philosophical argument, strictly speaking, consists mainly of an endeavour to cause the reader to perceive what has been perceived by the author. The argument, in short, is not of the nature of proof, but of exhortation' (quoted by Wood, 1975, p. 197). At the end of the argument we say, 'Look, can't you see what I see?' And so the debate goes on.

So much for the nature of philosophical reasoning, which is equally applicable to science and to religion. What we must avoid is the attempt to put our arguments beyond dispute by the irrational appeal to authority, whether the authority be of the Bible, the Leader, sense experience, the Scientist, intuition or revelation. The roots of the seductive appeal to authority go deep in epistemology, and it is a marked characteristic of man's search for understanding and knowledge. Popper identifies this desire for authority as the characteristic of all subjectivist epistemologies, and we have examined his philosophy in terms of a *Retreat from Authority*. It does not follow that we do not in practice accept certain things as authoritative; we cannot constantly question everything, otherwise like Buridan's Ass we would die of indecision. But we must always be prepared to submit these authorities to rational criticism and appraisal. We need to follow this course both in science and religion.

We have observed how positivism and relativism (the Received View and the *Weltanschauung* view of science), are linked through their common preoccupation with meaning. At a deeper level the link can be perceived in terms of the appeal to authority. The failure of positivism to establish an indubitable base for empirical science and to eradicate the metaphysical elements from scientific theories led to the elevation of the closed community as the final arbiter and source of authority, immune from external criticism. A parallel in European history is perceived in the Reformation which can be viewed as a switch of authority from the Church to the individual or eclectic enclave. The external standard or criterion of authority is replaced by the internal authority of the individual or collective *Weltanschauung*. This is the appeal of relativism which many in our day find attractive. In terms of knowledge and truth, one man's view is as good as another's − it is true for me. In terms of morality, there are no objective standards − it is right for me. In social and political terms, one man is as good as another − every man is his own master. In religion, each person or exclusive sect is secure and invulnerable from attack and from outside criticism − the individual soul before its Maker.

The attempt to provide an alternative to positivism, to scepticism, to relativism, made by Popper, has received most attention in this book. This is not only because his philosophy covers such a wide span, both in terms of time (half a century) and ideas, but because, for all the defects his critics have pointed to, it seems to be truer to the nature of man and to the world, and to offer the most satisfactory way forward in man's twofold enquiry in terms of his relationship with God and with the world. If the attempt to characterise science in clear and unequivocal terms has proved elusive, it is because science is a human activity; so is religion.

Notes

1. In his review of *The Concept of Mind* Russell wrote:

Professor Ryle's attitude to science is curious. He no doubt knows that scientists say things which they believe to be relevant to the problems he is discussing, but he is quite persuaded that the philosopher need pay no attention to science. He seems to believe that a philosopher need not know anything scientific beyond what was known in the time of our ancestors when they dyed themselves with woad. (1975, pp. 183-4)

Popper traces this disregard for science to the dubious influence of 'liberal education', and his remarks are of direct relevance to our argument:

in our day no man should be considered educated if he does not take an interest in science . . . For science is not merely a collection of facts . . . it is one of the most important spiritual movements of our day. Anybody who does not attempt to acquire an understanding of this movement cuts himself off from the most remarkable development in the history of human affairs. Our so-called Arts Faculties, based upon the theory that by means of a literary and historical education they introduce the student into the spiritual life of man, have therefore become obsolete in their present form. There can be no history of man which excludes a history of his intellectual struggles and achievements; there can be no history of ideas which excludes the history of scientific ideas. But literary education has an even more serious aspect. Not only does it fail to educate him to intellectual honesty. Only if the student an understanding of the greatest spiritual movement of his own day, but it also often fails to eduate him to intellectual honesty. Only if the student experiences how easy it is to err, and how hard to make even a small advance in the field of knowledge, only then can he obtain a feeling for the standards of intellectual honesty, a respect for truth, and a disregard for authority and bumptiousness. (1966, Vol. 2, note 6, p. 283)

2. Norman Malcolm suggests a number of reasons for this.

One may be the illusion that science can justify its own framework. Another is the fact that science is a vastly greater force in our culture. Still another may be the fact that by and large religion is to university people an alien form of life. They do not participate in it and do not understand what it is all about.
    Their nonunderstanding is of an interesting nature. It derives, at least in part, from the inclination of academics to suppose that their employment as

scholars demands of them the most severe objectivity and dispassionateness. For an academic philosopher to become a religious believer would be a stain on his professional competence! (1977, p. 156)

REFERENCES

Alexander, P. (1963) *Sensationalism and Scientific Explanation*, Routledge and Kegan Paul, London

Altizer, T. (1967) *The Gospel of Christian Atheism*, Collins, London

Ayer, A.J. (1946) *Language, Truth and Logic*, rev. edn, Gollancz, London. First published 1936

—— (1956) *The Problem of Knowledge*, Penguin Books, London

—— (1959) *Logical Positivism*, Allen and Unwin, London

—— (1971) 'Conversation with A.J. Ayer' in Magee (1971)

—— (1976) *The Central Questions of Philosophy*, Penguin Books, Harmondsworth. First published 1973

—— (1978) 'Logical Positivism and its Legacy' in Magee (1978)

—— (1979) 'Replies' in Macdonald (1979)

Baker, J.A. (1970) *The Foolishness of God*, Darton, Longman and Todd, London

Bambrough, R. (1977) 'Introduction' in Brown (1977)

Bartley, W.W. III (1962) *The Retreat to Commitment*, Knopf, New York

—— (1964) 'Rationality versus the Theory of Rationality' in Bunge (1964)

—— (1974) *Wittgenstein*, Quartet Books, London

Barton, J. (1979) 'Cultural Relativism', *Theology* (March and May), Society for Promoting Christian Knowledge, London

Berlin, I. (1968) 'Verification' in Parkinson (1968). First published 1938

Bhaskar, R. (1978) *A Realist Theory of Science*, Harvester Press, Hassocks, Sussex

Bondi, H. (1962) *Relativity and Common Sense: A New Approach to Einstein*, Heinemann, London

Braithwaite, R.B. (1953) *Scientific Explanation*, Harper Torchbooks, New York

—— (1966) 'Probability and Induction' in Mace (1966)

—— (1971) 'An Empiricist's View of the Nature of Religious Belief' in Mitchell (1971). First published 1955

Bridgman, P.W. (1927) *The Logic of Modern Physics*, Macmillan, New York

Brown, S.C. (ed.) (1977) *Reason and Religion*, Cornell University Press, London

Bultmann, R. (1958) *Jesus Christ and Mythology*, Charles Scribner's Sons, New York

Bunge, M.I. (ed.) (1964) *The Critical Approach to Science and Philosophy*, The Free Press of Glencoe, Collier-Macmillan, London

Buren, P. van (1963) *The Secular Meaning of the Gospel*, Student Christian Movement Press, London

Campbell, D.T. (1974) 'Evolutionary Epistemology' in Schilpp (1974)

Carnap, R. (1934) 'Psychology in Physical Language' in *The Unity of Science*, translated by M. Black, Kegan Paul, London

—— (1953) 'Testability and Meaning' in Feigl and Brodbeck (1953). First published 1936

Carroll, L. (1871) *Through the Looking Glass and What Alice Found There*. Currently published by Penguin Books, Harmondsworth

Collingwood, R.G. (1940) *An Essay on Metaphysics*, Clarendon Press, Oxford

Cupitt, D. (1977) 'The Christ of Christendom' in Hick (1977a)

—— (1979) *Jesus and the Gospel of God*, Lutterworth Press, London

—— (1980) *Taking Leave of God*, Student Christian Movement Press, London

—— (1983) 'Religion and Critical Thinking', *Theology* (July and September), Society for Promoting Christian Knowledge, London

Darwin, C. (1902) *The Origin of Species by Means of Natural Selection*, John Murray, London. First published 1859

Dewey, J. (1934) *A Common Faith*, Yale University Press, New Haven, Conn.

Eccles, J.C. (1974) 'The World of Objective Knowledge' in Schilpp (1974)

—— (1979) *The Human Mystery*, Springer International, Berlin

Edwards, D.L. (ed.) (1963) *The Honest to God Debate*, Student Christian Movement Press, London

—— (1969) *Religion and Change*, Hodder and Stoughton, London

Einstein, A. and Infeld, L. (1961) *The Evolution of Physics*, Cambridge University Press, Cambridge. First published 1938

Farrer, A. (1972) *Reflective Faith*, Society for Promoting Christian Knowledge, London

Feigl, H. and Brodbeck, M. (eds.) (1953) *Readings in the Philosophy of Science*, Appleton-Century-Crofts, New York

Feyerabend, P.K. (1975) *Against Method*, New Left Books, London

—— (1978) *Science in a Free Society*, New Left Books, London

—— (1981) *Philosophical Papers*, 2 vols., Cambridge University Press, Cambridge

Flew, A. (1966) *God and Philosophy*, Hutchinson, London

—— (1971) 'Theology and Falsification' in Mitchell (1971). First published 1955

Foster, M.B. (1973) 'The Christian Doctrine of Creation and the Rise of Modern Natural Science' in Russell (1973). First published 1934

Freeman, E. (ed.) (1976) *The Abdication of Philosophy: Philosophy and the Public Good*, Open Court, La Salle, Ill.

Green, M. (ed.) (1977) *The Truth of God Incarnate*, Hodder and Stoughton, London

Hacking, I. (ed.) (1981) *Scientific Revolutions*, Oxford University Press, Oxford

Hanson, N.R. (1958) *Patterns of Discovery*, Cambridge University Press, London

Hempel, C. (1965) 'Theoretician's Dilemma' in *Aspects of Scientific Explanation and Other Essays in the Philosophy of Science*, New York Free Press, New York. First published 1958

Hesse, M. (1961) *Forces and Fields: The Concept of Action at a Distance in the History of Physics*, Nelson, London

—— (1964) 'Francis Bacon' in O'Connor (1964)

Hick, J. (1971) 'Theology and Verification' in Mitchell (1971) First published 1960

—— (1973) *God and the Universe of Faiths*, Macmillan, London

—— (ed.) (1977a) *The Myth of God Incarnate*, Student Christian Movement Press, London

—— (1977b) 'Jesus and the World Religions' in Hick (1977a)

—— (1980) *God has Many Names*, Macmillan, London

Hoffmann, B. (1975) *Einstein*, Paladin, St Albans

Hookyaas, R. (1972) *Religion and the Rise of Modern Science*, Scottish Aca-

200 *References*

demic Press, Edinburgh
Hoyle, F. (1983) *The Intelligent Universe*, M. Joseph, London
Hudson, W.D. (1968) *Ludwig Wittgenstein*, Lutterworth Press, London
—— (1975) *Wittgenstein and Religious Belief*, Macmillan, London
Hume, D. (1951) *An Enquiry Concerning Human Understanding*, ed. D.C. Yalden-Thomas, Nelson and Sons, London. First published 1740
Huxley, J. (1931) *Science and Religion*, Gerald Howe, London
Jaki, S.L. (1978) *The Road of Science and the Ways to God*, Scottish Academic Press, Edinburgh
Körner, S. (1979) 'Ayer on Metaphysics' in Macdonald (1979)
Krige, J. (1980) *Science, Revolution and Discontinuity*, Harvester Press, Hassocks, Sussex
Kuhn, T.S. (1970a) *The Structure of Scientific Revolutions*, 2nd edn, University of Chicago Press, Chicago. First published 1962
—— (1970b) 'Logic of Discovery or Psychology of Research?' in Lakatos and Musgrave (1970). Also in Schilpp (1974)
—— (1970c) 'Reflections on my Critics' in Lakatos and Musgrave (1970)
—— (1977a) 'Second Thoughts on Paradigms' in Suppe (1977)
—— (1977b) *The Essential Tension: Selected Studies in Scientific Tradition and Change*, Chicago University Press, Chicago
Laing, R.D. and Esterson, A. (1970) *Sanity, Madness and the Family*, Penguin Books, Harmondsworth. First published 1964
Lakatos, I. (1970) 'Falsification and the Methodology of Scientific Research Programmes' in Lakatos and Musgrave (1970)
—— (1974) 'Popper on Demarcation and Induction' in Schilpp (1974)
—— and Musgrave, A. (eds.) (1968) *Problems in the Philosophy of Science*, North Holland, Amsterdam
—— (eds.) (1970) *Criticism and the Growth of Knowledge*, Cambridge University Press, London
Lewis, H.D. (1965) *Philosophy of Religion*, Teach Yourself Books, The English Universities Press, London
Lieberson, J. (1983) 'The Karl Popper Problem', *The New York Review of Books*, 28 April, A.Whitney Ellsworth, New York
Lucas, J.R. (1970) *The Freedom of the Will*, Clarendon Press, Oxford
Macdonald, G.F. (ed.) (1979) *Perception and Identity*, Macmillan, London
Mace, C.A. (ed.) (1966) *British Philosophy in the Mid Century*, Allen and Unwin, London. First published 1957
MacIntyre, A. (1963) 'God and the Theologians' in Edwards (1963)
Mackie, J.L. (1982) *The Miracle of Theism*, Clarendon Press, Oxford
MacKinnon, D. (1968) *Borderlands of Theology*, Lutterworth Press, London
Magee, B. (1971) *Modern British Philosophy*, Secker and Warburg, London
—— (1973) *Popper*, Fontana-Collins, London
—— (1978) *Men of Ideas – Some Creators of Modern Philosophy*, British Broadcasting Corporation, London
Malcolm, N. (1967) 'Anselm's Ontological Argument' in Phillips (1967). First published 1960
—— (1977) 'The Groundlessness of Belief' in Brown (1967)
Marcuse, H. (1978) 'Marcuse and the Frankfurt School' in Magee (1978)
Masterman, M. (1970) 'The Nature of a Paradigm' in Lakatos and Musgrave (1977)

Meynell, H. (1977) 'The Intelligibility of the Universe' in Brown (1977)

Mitchell, B. (ed.) (1971) *The Philosophy of Religion*, Oxford University Press, London

Monod, J. (1972) *Chance and Necessity*, Collins, London

Munévar, G. (1981) *Radical Knowledge*, Avebury, Amersham, Bucks

Munz, P. (1964) 'Popper and Wittgenstein' in Bunge (1964)

Nagel, E., Suppes, P. and Tarski, A. (eds.) (1962) *Logic, Methodology and Philosophy of Science of the 1960 International Congress*, Stanford University Press, Stanford, Calif.

Neill, S. (1977) 'Jesus and History' in Green (1977)

Neurath, M. and Cohen, R.S. (eds.) (1973) *Otto Neurath: Empiricism and Sociology*, D. Reidel, Dordrecht, Netherlands

Newton-Smith, W.H. (1981) *The Rationality of Science*, Routledge and Kegan Paul, London

Nicholson-Lord, D. (1983) 'The Mass Killings that Put Psychology on Trial', *The Times*, 5 November

Nineham, D. (1976) *The Use and Abuse of the Bible*, Macmillan, London

—— (1977) 'Epilogue' in Hick (1977a)

O'Connor, D.J. (ed.) (1964) *A Critical History of Western Philosophy*, Collier-Macmillan, London

O'Hear, A. (1980) *Karl Popper*, Routledge and Kegan Paul, London

Parkinson, G.H.R. (ed.) (1968) *The Theory of Meaning*, Oxford Readings in Philosophy, Oxford University Press, London

Passmore, J. (1957) *A Hundred Years of Philosophy*, Duckworth, London

Pears, D. (1971) 'Conversation with David Pears' in Magee (1971)

Pearson, K. (1936) *The Grammar of Science*, Everyman, J.M. Dent and Sons. First published 1892

Phillips, D.Z. (1965) *The Concept of Prayer*, Routledge and Kegan Paul, London

—— (ed.) (1967) *Religion and Understanding*, Basil Blackwell, Oxford

—— (1970a) *Faith and Philosophical Enquiry*, Routledge and Kegan Paul, London

—— (1970b) *Death and Immortality*, Macmillan, London

—— (1976) *Religion without Explanation*, Basil Blackwell, Oxford

Polanyi, M. (1957) *Personal Knowledge*, Routledge and Kegan Paul, London

Popper, K.R. (1934) *Logik der Forschung*, J. Springer, Vienna

—— (1957) *The Poverty of Historicism*, Routledge and Kegan Paul, London

—— (1966) *The Open Society and its Enemies*, 2 vols., Routledge and Kegan Paul, London

—— (1968a) *The Logic of Scientific Discovery*, Hutchinson, London. First published 1959. Expanded version of *Logik der Forschung*, 1934

—— (1968b) 'Remarks on the Problems of Demarcation and Rationality' in Lakatos and Musgrave (1968)

—— (1970) 'Normal Science and its Dangers' in Lakatos and Musgrave (1970)

—— (1971) 'Conversation with Karl Popper' in Magee (1971)

—— (1972) *Conjectures and Refutations*, Routledge and Kegan Paul, London. First published 1963

—— (1974) 'Replies to my Critics' in Schilpp (1974)

—— (1976a) *Unended Quest*, Fontana, London. First published as 'Intellectual Autobiography' in Schilpp (1974)

—— (1976b) 'The Myth of the Framework' in Freeman (1976)

—— (1979) *Objective Knowledge*, Oxford University Press, Oxford. First published 1972

—— (1981) 'The Rationality of Scientific Revolutions' in Hacking (1971)

—— (1982a) *Quantum Theory and the Schism in Physics*, Hutchinson, London

—— (1982b) *The Open Universe*, Hutchinson, London

—— (1983) *Realism and the Aim of Science*, Hutchinson, London

—— and Eccles, J.C. (1977) *The Self and its Brain*, Springer International, Berlin

Putnam, H. (1962) 'What Theories are Not' in Nagel, Suppes and Tarski (1962)

—— (1974) 'The Corroboration of Theories' in Schilpp (1974)

—— (1978) 'The Philosophy of Science' in Magee (1978)

Quine, W.V.O. (1953) *From a Logical Point of View*, Harvard University Press, Cambridge, Mass.

—— (1974) 'On Popper's Negative Methodology' in Schilpp (1974)

Quinton, A.M. (1964) 'Contemporary British Philosophy' in O'Connor (1964)

—— (1982) *Thoughts and Thinkers*, Duckworth, London

Reichenbach, H. (1938) *Experience and Prediction*, University of Chicago Press, Chicago

Richmond, J. (1980) 'Religion without Explanation', *Theology* (January), Society for Promoting Christian Knowledge, London

Robinson, J.A.T. (1963) *Honest to God*, Student Christian Movement Press, London

Rucker, R. (1982) *Infinity and the Mind*, Harvester Press, Hassocks, Sussex

Russell, B. (1929) *Mysticism and Logic*, Allen and Unwin, London. First published 1917

—— (1946) *History of Western Philosophy*, Allen and Unwin, London

—— (1952) *The Problems of Philosophy*, Home University Library, Oxford University Press, London. First published 1912

—— (1975) *My Philosophical Development*, Allen and Unwin, London. First published 1959

Russell, C.A. (ed.) (1973) *Science and Religious Belief*, University of London Press, London

Ryle, G. (1954) *Dilemmas*, Cambridge University Press, Cambridge

Saifullah Kahn, V. (1976) 'Perceptions of a Population: Pakistanis in Britain', *New Community*, 5 (3) (Autumn)

Schilpp, P.A. (ed.) (1949) *Albert Einstein: Philosopher-Scientist*, Open Court, La Salle, Ill.

—— (ed.) (1974) *The Philosophy of Karl Popper*, 2 vols., Open Court, La Salle, Ill.

Shapere, D. (1982) 'Meaning and Scientific Change' in Hacking (1981). First published 1966

Stove, D.C. (1981) *Popper and After*, Pergamon Press, Oxford

Strawson, P.F. (1959) *Individuals. An Essay in Descriptive Metaphysics*, Methuen, London

Suppe, F. (ed.) (1977) *The Structure of Scientific Theories*, University of Illinois Press, Chicago

Swinburne, R. (1977) *The Coherence of Theism*, Clarendon Press, Oxford

—— (1979) *The Existence of God*, Clarendon Press, Oxford

Toulmin, S. (1953) *The Philosophy of Science: An Introduction*, Hutchinson, London
—— (1961) *Foresight and Understanding*, Hutchinson, London
Trigg, R. (1973) *Reason and Commitment*, Cambridge University Press, London
Vanstone, W.H. (1977) *Love's Endeavour, Love's Expense*, Darton, Longman and Todd, London
—— (1982) *The Stature of Waiting*, Darton, Longman and Todd, London
Ward, K. (1982a) *Holding Fast to God: A Reply to Don Cupitt*, Society for Promoting Christian Knowledge, London
—— (1982b) *Rational Theology and the Creativity of God*, Basil Blackwell, Oxford
Watkins, J.W.N. (1970) 'Against Normal Science' in Lakatos and Musgrave (1970)
—— (1974) 'The Unity of Popper's Thought' in Schilpp (1974)
Watson, J. (1968) *The Double Helix: A Personal Account of the Structure of D.N.A.*, Weidenfeld and Nicolson, London
Whitehead, A.N. (1938) *Science and the Modern World*, Penguin Books, London. First published 1926
—— and Russell, B. (1910-13) *Principia Mathematica*, 3 vols., Cambridge University Press, Cambridge
Wiles, M. (1977) 'Christianity without Incarnation' in Hick (1977a)
Williams, H.A. (1972) *True Resurrection*, Mitchell Beazley, London
Winch, P. (1958) *The Idea of a Social Science*, Routledge and Kegan Paul, London
—— (1967) 'Understanding a Primitive Society' in Phillips (1967). First published 1964
Wisdom, J. (1953) 'Gods' in *Philosophy and Psychoanalysis*, Basil Blackwell, Oxford
Wittgenstein, L. (1951) *Tractatus Logico-Philosophicus*, Routledge and Kegan Paul, London. First published 1922
—— (1953) *Philosophical Investigations*, translated by G.E.M. Anscombe, Basil Blackwell, Oxford
—— (1966) 'Lectures on Religious Belief' in C. Barrett (ed.), *Lectures and Conversations on Aesthetics, Psychology and Religious Belief*, Basil Blackwell, Oxford
Wood, A. (1975) 'Russell's Philosophy: A Study of its Development' in Russell (1975). First published 1959

INDEX

ad hoc hypotheses 89
Adler, A. 7, 67, 74, 133
Alexander, P. 11-12
Altizer, T. 42
analytic justification 84
analytic-synthetic distinction 24, 32-3
Anaxagoras 1
Anselm 174, 180
Aquinas, T. Preface, 180
Aristotle 6, 7, 25, 110
atheism 37, 46, 174
Augustine 123, 180
authoritarianism 50, 116-20
authority 117-18; of knowledge 4, 71; of science Preface, 1, 4, 6, 8, 46n1; of the senses 6-7, 35, 38
auxiliary hypotheses 89
Ayer, A.J. 23, 31; and logical positivism 10, 47n10, 48-9, 69; and phenomenalism 16-17, 84; and religion 38, 102, 181; and Verification Principle 16, 25-30, 46n6, 47n11, 70, 177

Bacon, F. 35, 110; and empiricism 6-8, 10, 12, 24, 47n9; and induction 20, 22-3
Baker, J.A. 194
Bambrough, R. 178
Barth, K. 36-7, 180
Bartley, W.W. III 15, 114, 169
Barton, J. 187n14
basic statements 87, 90
Beethoven, L. van 125
behaviourism 53
Berkeley, G. 12, 70, 80
Berlin, I. 28, 116
Bhasker, R. 30, 181-2
Bible 73, 110-11, 117
Bohr, N. 77, 90
Bondi, H. 134n1, 186n7
Bonhoeffer, D. 40
Bradley, F.H. 9
Braithwaite, R.B. 23, 38-40, 47n15, 180
Bridgman, P.W. 19
Buchner, L. 10

bucket theory of mind 80
Buhler, K. 61
Bultmann, R. 36, 47n13
Bunyan, J. 39
Buren, P van 40-2, 180

Calvin, J. 123
Campbell, D.T. 51-2, 63-5, 84, 134n7
Cantor, G. 13
Carnap, R. 16, 22, 29, 32-3, 47n12, 53
Carroll, L. 152
category mistake 163
catholicism 154
certainty 71
Chalcedonian formula 44
chance and necessity 55
Chardin, T. de Preface
Christian Church 117-18, 132
Christianity 168; and politics 115; and reductionism 38-42, 182
Christology 43-4
Churchill, W. 80
Collingwood, R.G. 25
Compton's problem 64, 72
Comte, A. 8
condition of meaning invariance 149
confirmation of theories 22-3, 34
conjecture and refutation 67-9, 82, 98, 133
consciousness 56, 60
consistency condition 149
context of discovery 136, 161, 186n1
Copernican revolution 1, 143
Copernicus, N. 2, 11, 77, 128
correspondence rules 13, 18, 32-3, 141
corroboration 94-5, 99
creation, Christian doctrine of 101, 128-32
Crick, B. and Watson, J. 54
criticism 4; and growth of knowledge 63, 65, 72-3, 76, 79, 88; and rationality 50, 72, 92, 108, 113-14, 116, 133, 193, 195; and truth 90, 113; Feyerabend on 150-2; in

metaphysics and religion 103-4,
111-13, 193; in politics 116;
objective 72
Cupitt, D. 43-4, 47n18, 187n16

Darwin, C. 95, 127-8, 150; and
natural selection 52, 53-5, 74, 79
Darwinism 7, 58, 103
decision making 68, 87
deductive logic 20, 84
definition 18
demarcation; between science and
non-science 49, 69-70, 74-9, 92,
102-8, 132-3; between sense and
nonsense 24, 30, 45; within metaphysics 105-6, 193-4
democracy 117
Democritus 54
Descartes, R. 45, 60, 71-2, 110
descriptive epistemology 51-3, 56,
80, 100, 132, 184-5
determinism 53, 105, 119; scientific
and metaphysical 122-7
Dewey, J. 39
disciplinary matrixes; *see* Kuhn
Dostoevesky, F. 39

Eccles, J.C. 56, 102, 112, 128,
134n1, 134n4
Eddington, A. 74, 101
Edwards, D. 37
Einstein, A. 1, 74, 104, 141, 146,
150, 184, 191-2; and creative
imagination 83, 126, 134n11;
and falsification 67-8, 77-9, 94;
and Infeld, L. 66, 182, 186n7;
and positivism 12
emergence; of consciousness 53, 56;
of language 59-62, 72
empirical tradition 6-10
empiricism 6-10, 22, 23-4, 27, 35,
80, 149
epiphenomenalism 56
epistemic fallacy 30
epistemological anarchy 149, 152-3,
155
epistemological atomism 11
epistemology 2, 11, 70-1; and society
Preface, 110, 115, 155;
traditional 63, 70-1
eschatological verification 42-3
Essentialism 67, 122, 152, 159, 162
evolution; *see* Popper
evolutionary continuum 52-3

evolutionary epistemology 51, 63,
65-7, 69, 76, 79, 134
exemplars; *see* Kuhn
explanatory myths 76
extensionality 14

faith 112
fallibilism Preface, 23; in Popper 50,
73, 84, 88, 99, 109, 191; in
religion 112-13
falsification and demarcation 76; and
indeterminism 126; and truth 77,
90; logical 84, 85, 135n13, 136;
methodological 86-90, 136; of
religious assertions 38; of theories
50, 66-7, 69, 75, 88-90, 149
Faraday, M. 104
Farrer, A. 112
Feyerabend, P.K. and relativism 87,
137, 158-9, 161, 182, 191; and
weltanschauung 137, 154, 156,
160, 172, 181; and Wittgenstein
161-2, 164-5, 166-7; critique of
154-5; irrationality of science 77,
149-52, 183-4; science and
society 152-4
fideism 106, 170, 178, 182
Flew, A. 38, 40, 42
form of life *see* Wittgenstein
Frege, G. 13-14, 46
Freud, S. 7, 67, 74, 79, 134n10
fundamentalism 111, 188n17

Galileo 1, 47n9, 77, 152
Garbo, G. 53
Gellner, E. 169
genetic dualism 55-6
Gibbon, E. 44
God 1, 115, 119, 189-90, 193-4;
and determinism 123; and
evolution 127; existence of 3, 25,
27, 38, 42-3, 100; *see also* Barth;
van Buren; Hick; Phillips;
Vanstone
Godel, K. 125, 135n24
God-talk 26, 40-5, 193
Goodman, N. 17
Greenslade, T.A. Preface
Grisez, G. 30

Hanson, N.R. 33, 87, 137
Harnack, A. 36
Hawking, S. 101
Heisenberg, W. 143

Hempel, C. 19
Heraclitus 103
Herodotus 44
Hesse, M. 7, 21
Hick, J. 42-3, 45, 106-7
Hiroshima 126
historicism 116, 118-19
history 44, 78, 119, 138, 195
Hitler, A. 116
Hobbes, J. 124
holistic reforms 120
Homer 125
Hookyaas, R. 5n2
Hoyle, F. 124n5
Hudson, W.D. 165, 169
human freedom and creativity 122-7
Hume, D. and deduction 84; and
    determinism 55, 124; and
    empiricism 2, 12; and idealism 80;
    and induction 21-3, 49, 80-2,
    85, 94, 98; and metaphysics
    25-6, 176; and natural theology
    173-4; and scepticism 99
Huxley, J. 113
Huxley, T. 127

idealism 9, 80, 105, 151, 157, 161
identity theory 56
immunizing stratagems 89
inborn expectations 83-4
incarnation, Christian doctrine of
    43-5
incommensurability of theories
    145-6, 148, 150-1, 157, 161
indeterminism 55-7, 69, 92, 104,
    122-7
induction 29, 34, 52, 133; *see*
    Bacon; Hume, Popper
instrumentalism 18-19
Ionian philosophers 1
Iran 120
irrationalism 105, 110, 148, 151
Islam 115, 168

Jaki, S.L. 5n2
Jesus 41, 43-4, 112
Johnson, S. 80
Judaism 115, 168
Jung, C. 7
justification 21, 35, 50, 71, 73, 81,
    99

Kant, I. 11, 111, 156-7, 169;
    analytic-synthetic distinction 24;

and determinism 124; and meta-
    physics 25
Keller, H. 60
Keppler, J. 2, 77
Khan, S. 167
Kierkegaard, S. 174
knowledge; commonsense theory
    70-1, 79-80; conjectural 68, 80;
    growth of 49-51, 66, 69, 76, 88;
    inborn 60, 111; scientific 2, 7-9,
    11, 15, 20-3, 26, 66, 70, 79, 93,
    145, 147; subjective and objective
    70-2; theological 9, 112
Körner, S. 31
Krige, J. 89, 155, 183-4, 181n1,
    186n7
Kuhn, T. 49, 52, 86, 99, 137, 165,
    173; and relativism 87, 158-9,
    161; and science, crisis in 143-5;
    criticism in 77-9; history of
    138-9, 147, 162, normal 139,
    141-3, 145, 147-8, 151, progress
    in 145-6, view of 137-8, 179, 181;
    and scientific community 141-2,
    144-8, 176, 191; and *weltan-
    schauung* 137, 139-42, 146, 148,
    150, 156, 160-1, 164, 172, 181;
    disciplinary matrix 140-8;
    exemplars 140-1, 171; paradigms
    135n15, 138-41, 148, 157, 162,
    166-7

Laing, R. and Esterson, A. 62
Lakatos, I. 79, 86, 89, 135n15, 152;
    and induction 98-9, 133
Lamarck, J. 51, 54, 65, 67, 74
Lamb, D. Preface
language 17, 159; evolution of
    59-63, 72; functions of 61-3, 71,
    74, 185; learning 63-4; *see also*
    Wittgenstein
Laplace, P. 102, 122
Last Judgement 170, 177
Lewis, C.I. 17
Lewis, H.D. 102, 109
liberal protestant theology 35-7, 43
Lieberson, J. 96-7
linguistic analysis 40-1
Locke, J. 2, 7, 12, 52, 70-1, 83
logic; inductive 134n12, 136; of
    discovery 98, 136-7; of the
    situation 79, 84, 86, 90, 133
logical: analysis 13-14, 50;
    empiricism 22; necessity 23;

positivism 9-10, 13, 16, 31, 45, 50
Logical Positivists 12, 17-19, 76, 87; and induction 23; and rationality 14, 22; and reductionism 10; and *Tractatus* 162, 169
love 111, 113-14; of God 128-32
Lucas, J. 125
Luther, M. 123

Mach, E. 11-12, 18, 27, 80
MacIntyre, A. 37
Mackie, J.L. 193
MacKinnon, D. 36-7
McTaggart, J. 9
Magee, B. 48, 102, 113, 116, 120, 162
Malcolm, N. 172, 174, 196n2
Marcuse, H. 135n22
Marx, K. 7, 67, 74-5, 117, 119, 126, 134n10
Marxism 116, 118-20
Masterman, M. 140
mathematics 13-14
Maxwell, J.C. 142
meaning: theory-dependence of 148-9, 151, 156-61; *see also* Wittgenstein
mechanistic materialism 10-11
Medawar, P. 134n1
Mendel, G. 54, 95
metalanguages 92
metaphysical: beliefs 25-7, 85 research programmes 55, 103, 135n15
metaphysics: and demarcation 69, 76, 90, 102-8; and positivism 25, 41, 100; and religion 2, 8, 10, 24-8; primary and secondary systems 31; views of Aristotle, Collingwood, Kant, Strawson 25
Meynell, H. 30
Michaelangelo 126
Mill, J.S. 8, 153
mind-body dualism 56
Modernism 2, 35
modus tollens 86, 89, 136
Monod, J. 54-5, 59
Moore, G. 9
Morrell, Lady O. 80
Mozart, W. 124, 126, 128
Munevar, G. 134n7, 184
Munz, P. 170, 187n12
mutations 56-7

Napoleon 102
natural selection 53-4
natural theology 36, 173-4
Neill, S. 112
neo-Darwinism 54, 65
Neptune 89
Neurath, M. and Cohen, R.S. 10, 18
neutrino 89
New Testament 36, 43
Newton, I. 1-2, 101, 109; and Einstein 68, 77, 141, 145-6, 149-50; and induction 21; theory of gravitation 79, 89, 94
Newton-Smith, W.H. 97
Nicholson, Lord, D. 133
nihilism 105
Nilsen, D. 133
Nineham, D. 44, 187n14
normal science *see* Kuhn

objectivism 104
observation 13, 78; theory dependence of 50, 68, 83, 87, 136, 148-9, 151, 156-7, 161
observational-theoretical distinction 32-3
O'Hear, A. 88, 96-7, 114
open society *see* Popper, political philosophy
operationalism 19

parable of the gardener 38
paradigms *see* Kuhn
Parmenides 103
Passmore, J. 11
Pauli, W. 89, 143
Pears, D. 162
Pearson, K. 20
Perrin, N. 44
persuasive definitions 31
phenomenalism 16-17, 29-31
Phillips, D.Z. 172; and natural theology 173-4; and religious language 194; meaning and existence 178-83; philosophy of religion 173-8
philosophy: of physics 8; of politics 115-21; of religion Preface, 4, 9, 51, 132, 170, 173-8, 191; of science Preface, 3-4, 8-10, 12-13, 23, 34, 49, 52-3, 77-8, 124, 115-16, 136, 191; task of Preface, 2, 161, 169, 173, 189

physicalism 16, 22, 53
physics 8, 12, 32, 66-7, 103
Planck, M. 12
Plant, R. Preface
plastic control 56, 126
Plato 7, 70, 79, 86, 117, 125, 161
Polyanyi, M. 137
Popper, K.R. Preface, 2, 32, 46, 139,
    153, 155-6; and Wittgenstein
    165, 170, 187n12; essentialism
    67, 122, 152, 159; evolution of
    knowledge, conjecture and
    refutation 67-9, demarcation
    69-70, 74-9, evolutionary
    epistemology 65-7, fallibilism 73,
    191, induction 67, 76-7, 79-90,
    96-9, subjective and objective
    knowledge 70-2, truth 90-3,
    109-111, 145, 184-5, verisimili-
    tude 93-6, world 3, 72-3, 109;
    evolution of persons, biological
    evolution 56-9, consciousness
    56, Darwinism 53-5, emergence
    and reduction 53, indeterminacy
    55, language 59-65, linguistic
    subjectivism 63, plastic controls
    64, problem solving 7, 56-9;
    implications for religion,
    authority 108-115, 195, evolution
    127-32, indeterminism 122-7,
    metaphysics and demarcation
    102-8, 193-4, political philo-
    sophy 72, 115-21, spirit of
    enquiry 99-102; influence and
    range of ideas 48-52; liberal
    education 196n1; relativism 172
positivism Preface; and empiricism
    6-10, 22; and Hume 80, 173; and
    Kuhn 148; and metaphysics 24,
    103, 190-1; and probability 93;
    and theology 2, 4, 40, 42; and the
    Received View of science 10-13;
    and relativism 182; Christological
    36-7; demise of 31-5, 86
prediction 118-19
pre-science 139
principle of transference 81
probability 93, 109
problem solving 50, 56, 63, 116, 169
propensity theory 104
psychological primacy of repetitions
    81-3
psychology 7
Putnam, H. 33, 35, 46n3, 97, 99

Pythagoras 103

qualia 17, 84
quantum mechanics (and theory) 7,
    12, 90, 103-4, 124, 142-3, 149
Quine, W. 32-3, 89, 172
Quinton, A. 5n4, 14

Ramsey, F. 19
rationality 114; and criticism 50,
    72-4, 92-3, 108, 113-14, 116,
    133, 195; and induction 23; and
    verification 43; Hume's view 22;
    in religion 27, 193; in science 2,
    12, 14, 37, 77, 85, 96, 98-9, 153;
    in science and metaphysics 1,
    3, 106-8, 192
realism 69, 80, 91, 104, 181-2, 184;
    see also Feyerabend
Received View on theories 46n3,
    109, 138, 158, 162, 195; and
    determinism 122; and discovery
    136-7; and religion 100; demise
    of 31-5, 137, 141, 156, 176, 183;
    development of 10-14, 17-20,
    23-4, 29, 33
reduction(ism); and determinism 53,
    123; in religion 2, 39-41, 106; in
    science 10
Reichenbach, H. 12, 186n1
relativism 4; and religion 106, 183,
    195; of Kuhn and Feyerabend 87,
    137-8, 151, 158-9, 162, 171;
    Popper on 9, 110, 172
relativity theory 7, 12, 74, 125, 139,
    145, 149
religion; and politics 115; and
    positivism 10, 35, 105; and
    science Preface, 1-3, 6, 9, 27,
    35-6, 100, 102, 111; and theology
    4, 5n1; philosophy of: *see*
    philosophy
religious: belief 39, *see also* Phillips;
    Wittgenstein; communities 147-8,
    167-9; language 2, 100; as expres-
    sive 177-9; pictures 171; pluralism
    154
resurrection 42, 179
revelation 36
revolutions in science 147
Rhees, R. 172, 177
Richmond, J. 182
Robinson, J. 42
Rucker, R. 135n24

Russell, B. 2, 50, 70, 80, 110, 189,
    191-2, 194; and analysis 13-14,
    36, 46n4, 51; and induction 22-3,
    79, 85-6; and language 159, 185;
    and phenomenalism 16; and
    Wittgenstein 52, 169, 170,
    187n12
Russell's paradox 46, 92
Rutherford, E. 90, 101
Ryle, G. 29, 163, 196n1

Saint John the Divine 190, 192
Saint Paul 44-5, 180
Sands, B. 44
Santayana, G. 97
scepticism 99
Schilpp, P. 7, 34n1, 192
schizophrenia 62, 84
Schlick, M. 12, 19, 24, 42
Schrödinger, E. 104
Schweitzer, A. 36
science *see* Feyerabend; Kuhn;
    Popper on evolution of know-
    ledge, indeterminism, meta-
    physics and demarcation; *see also*
    philosophy of science, religion;
    theories
scientific community *see* Kuhn;
    critique of 167-9; knowledge *see*
    knowledge; research programmes
    135n15
searchlight theory of mind 80
sensationalism 11-12, 18
sense data 16-17, 84
sense experience 7, 24, 84
Shapere, D. 140, 149, 158-60
solipsism 80
Spencer, H. 134n7
Spinoza, B. 124, 169
Stalin, J. 116
stipulative definition 30
Stove, D.C. 152
Strawson, P.F. 25
Suppe, F. 7, 10, 13, 17, 31-5,
    46nn2&3, 140, 157, 161
Sutcliffe, P. 133
Swinburne, R. 193

Tarski, A. 90-2, 133
test statements 68, 87-8
theology 2, 8, 26-7, 41, 112; and
    evolution 128-32; as grammar
    164, 171, 178; Christian 1, 36-7,
    123-7; natural 5n2, 6, 173-4; *see*

also religion
Theoretician's Dilemma 19
theories; and decisions 87; and
    evolution 7, 58, 84; and meaning
    157-61; incommensurability of
    150, 157-8; of non-science 74-5,
    103-8; of science 71, 78, 82-3,
    93, 96, 136, 145, and the
    Received View 13-14, 18-20,
    31-4; social and political 16-17
Thucydides 44
totalitarianism 117
Toulmin, S. 19, 134n7, 137
Toynbee, A. 44
tradition 111
Trigg, R. 166, 168, 180
truth 183-5; and criticism 110-13;
    and knowledge 70-1; and religion
    42, 105-9; and verisimilitude 69,
    77, 93-6, 98, 149; as consensus
    146, 151; biological origin 62;
    correspondence theory 90-2,
    133-4, 156; regulative idea 62,
    93, 110, 145;
tu quoque 114

Unified Field Theory 78
Uranus 89
utopianism 119-21

validity 62
Vanstone, W.H. 128-32
verification; and falsification 50,
    67, 90; and knowledge 17, 19,
    26, 35, 42-3, 107; and meaning
    19, 24, 41; and Received View
    13, 34, 69; strong and weak 29
Verification Criterion 16, 19, 41, 69,
    76
Verification Principle 16, 18, 24, 30,
    42, 44, 48, 177; and religious
    statements 38-40, 106; failure of
    28-31
Verification Theory of Meaning
    13-18, 22
verisimilitude 69, 90-6, 139
Vienna Circle 7, 9-10, 15-16, 18, 45,
    50
voluntarism 105

Waismann, F. 19, 164
Ward, K. 47n18, 35n18
Watkins, J.W.N. 51, 125-7, 133, 147-8
Watson, W.H. 173

Weltanschauung(en) 136, 173, 182,
   192, 195; and Wittgenstein 162,
   165, 170; critique of 156-61,
   178; *see also* Feyerabend; Kuhn
Whitehead, A.N. 13, 5n2, 92, 101
Wilberforce, S. 127
Wiles, M. 43
Williams, H.A. 179
Winch, P. 119, 165-6, 172
Wisdom, J. 38
Wittgenstein, L.; and religious belief
   2, 100, 170-82; and *Tractatus*
   14-15, 24-5, 30, 42, 46n4, 52, 91,
   158, 161-5, 169-71; form of life
   119, 162-9, 170-1, 174, 176, 178,
   181, 194; language and reality 61,
   88, 164-7, 184-5, 191; language-
   games 162-6, 170-1, 174, 176,
   178, 180; *Philosophical Investigations* 161-4, 166, 169, 171-2
Wood, A. 191, 194-5

Zenophanes 109